ECOLOGICAL DYNAMICS
ON YELLOWSTONE'S NORTHERN RANGE

Committee on Ungulate Management in Yellowstone National Park

Board on Environmental Studies and Toxicology

Division on Earth and Life Studies

National Research Council

NATIONAL ACADEMY PRESS
Washington, D.C.

NATIONAL ACADEMY PRESS   2101 Constitution Ave., N.W.   Washington, D.C. 20418

NOTICE: The project that is the subject of this report was approved by the Governing Board of the National Research Council, whose members are drawn from the councils of the National Academy of Sciences, the National Academy of Engineering, and the Institute of Medicine. The members of the committee responsible for the report were chosen for their special competences and with regard for appropriate balance.

This project was supported by Cooperative Agreement 1443CX2605-98-005, between the National Academy of Sciences and the U.S. Department of Interior National Park Service. Any opinions, findings, conclusions, or recommendations expressed in this publication are those of the author(s) and do not necessarily reflect the view of the organizations or agencies that provided support for this project.

Library of Congress Control Number 2002108821

International Standard Book Number 0-309-083451

Additional copies of this report are available from:

National Academy Press
2101 Constitution Ave., NW
Box 285
Washington, DC 20055

800-624-6242
202-334-3313 (in the Washington metropolitan area)
http://www.nap.edu

THE NATIONAL ACADEMIES

National Academy of Sciences
National Academy of Engineering
Institute of Medicine
National Research Council

The **National Academy of Sciences** is a private, nonprofit, self-perpetuating society of distinguished scholars engaged in scientific and engineering research, dedicated to the furtherance of science and technology and to their use for the general welfare. Upon the authority of the charter granted to it by the Congress in 1863, the Academy has a mandate that requires it to advise the federal government on scientific and technical matters. Dr. Bruce M. Alberts is president of the National Academy of Sciences.

The **National Academy of Engineering** was established in 1964, under the charter of the National Academy of Sciences, as a parallel organization of outstanding engineers. It is autonomous in its administration and in the selection of its members, sharing with the National Academy of Sciences the responsibility for advising the federal government. The National Academy of Engineering also sponsors engineering programs aimed at meeting national needs, encourages education and research, and recognizes the superior achievements of engineers. Dr. Wm. A. Wulf is president of the National Academy of Engineering.

The **Institute of Medicine** was established in 1970 by the National Academy of Sciences to secure the services of eminent members of appropriate professions in the examination of policy matters pertaining to the health of the public. The Institute acts under the responsibility given to the National Academy of Sciences by its congressional charter to be an adviser to the federal government and, upon its own initiative, to identify issues of medical care, research, and education. Dr. Kenneth I. Shine is president of the Institute of Medicine.

The **National Research Council** was organized by the National Academy of Sciences in 1916 to associate the broad community of science and technology with the Academy's purposes of furthering knowledge and advising the federal government. Functioning in accordance with general policies determined by the Academy, the Council has become the principal operating agency of both the National Academy of Sciences and the National Academy of Engineering in providing services to the government, the public, and the scientific and engineering communities. The Council is administered jointly by both Academies and the Institute of Medicine. Dr. Bruce M. Alberts and Dr. Wm. A. Wulf are chairman and vice chairman, respectively, of the National Research Council.

COMMITTEE ON UNGULATE MANAGEMENT IN YELLOWSTONE NATIONAL PARK

Members

DAVID R. KLEIN *(Chair)*, University of Alaska Fairbanks, Fairbanks
DALE R. MCCULLOUGH *(Vice Chair)*, University of California, Berkeley
BARBARA ALLEN-DIAZ, University of California, Berkeley
NORMAN CHEVILLE, Iowa State University, Ames
RUSSELL W. GRAHAM, Denver Museum of Natural History, Denver, Colorado
JOHN E. GROSS, Commonwealth Scientific and Industrial Research Organization, Australia
JAMES MACMAHON, Utah State University, Logan
NANCY E. MATHEWS, University of Wisconsin, Madison
DUNCAN T. PATTEN, Montana State University, Bozeman
KATHERINE RALLS, Smithsonian Institution, Washington, D.C.
MONICA G. TURNER, University of Wisconsin, Madison
ELIZABETH S. WILLIAMS, University of Wyoming, Laramie

Project Staff

DAVID J. POLICANSKY, Project Director
LEE PAULSON, Staff Officer
MARGARET WALSH, Postdoctoral Research Associate
CAY BUTLER, Editor
KELLY CLARK, Editorial Assistant
MIRSADA KARALIC-LONCAREVIC, Research Assistant
RAMYA CHARI, Project Assistant
KATHY IVERSON, Senior Project Assistant

Sponsor

U.S. DEPARTMENT OF INTERIOR

Acknowledgment of Review Participants

This report has been reviewed in draft form by individuals chosen for their diverse perspectives and technical expertise, in accordance with procedures approved by the NRC's Report Review Committee. The purpose of this independent review is to provide candid and critical comments that will assist the institution in making its published report as sound as possible and to ensure that the report meets institutional standards for objectivity, evidence, and responsiveness to the study charge. The review comments and draft manuscript remain confidential to protect the integrity of the deliberative process. We wish to thank the following individuals for their review of this report:

Donald Barry, The Wilderness Society
Peter A. Bisson, U.S. Forest Service
Mark S. Boyce, University of Alberta
Ingrid Burke, Colorado State University
Charles C. Capen, Ohio State University
Michael B. Coughenour, Colorado State University
Amy E. Hessl, West Virginia University
Douglas B. Houston, National Park Service (retired)
Peggy Johnson, Pennsylvania State University

Walter Klippel, University of Tennessee
Frederic Wagner, Utah State University
James E. Womack, Texas A&M University

Although the reviewers listed above have provided many constructive comments and suggestions, they were not asked to endorse the conclusions or recommendations nor did they see the final draft of the report before its release. The review of this report was overseen by Simon Levin, Princeton University, and Ellis Cowling, North Carolina State University. Appointed by the National Research Council, they were responsible for making certain that an independent examination of this report was carried out in accordance with institutional procedures and that all review comments were carefully considered. Responsibility for the final content of this report rests entirely with the authoring committee and the institution.

Preface

National Park policy for management of ungulates within Yellowstone National Park (YNP) has undergone major changes since Yellowstone, the world's first national park, was established in 1872. These changes, from little emphasis on wildlife at the time of establishment to a major focus on wildlife today, have accompanied an evolving interest in wildlife by the visiting public together with increased understanding of the ecosystem relationships of wildlife by the National Park Service. The ecological relationships of the elk, bison, and other ungulates that inhabit the northern portion of YNP are, however, extremely complex. Acquiring adequate knowledge of this complexity as a basis for park management policy has been complicated by both the dynamic changes that have characterized the system and the recognition that the natural boundaries of this ecosystem extend well outside of YNP, largely to the north into Montana. As a consequence, any management activities that may affect the ungulates within YNP, such as previous reduction hunts and the more recent reintroduction of wolves to the ecosystem, are not independent of land use and associated human activities in the northern range area outside of the park.

The national park concept originated in the United States with the National Park Service's mandate for management of YNP has been to assure protection of the geological, landscape, and biological features for which it was established and to provide opportunities for the public to visit and appreciate its unique natural values. Interests and expectations of visitors to YNP regarding wildlife, although changing over time, have continued to influence wildlife management policy within the park. An understanding of the ecological relationships of wildlife within the park is also an important component of national park management. A major challenge for park management is to meet expectations of the visiting public regarding wildlife, with the cumulative increase in understanding of the ecology of park wildlife.

Within this context of ecosystem change and complexity, changing human perspectives and interests in wildlife, and markedly differing land designations and uses inside and outside of YNP, it is not surprising that controversy arose over National Park Service policy toward ungulates within the park. The National Research Council's Committee on Ungulate Management in Yellowstone National Park, charged to review information of the population ecology about ungulates on the northern range of the Greater Yellowstone ecosystem and associated practices for their management, was cognizant of the diversity of perspectives, concerns, and opinions that the public has expressed about management policy for ungulates in YNP. The committee appreciated the opportunity to hear from a wide range of people interested in the issues the committee was asked to address during open forums in Gardiner, Montana, and in Mammoth, Wyoming. The oral and written submissions, reports, and publications provided to the committee by members of the public; the National Park Service; the U.S. Forest Service; Montana Fish, Wildlife and Parks; the Biological Resources Division of the U.S. Geological Survey; and other agency representatives, were particularly valuable in our efforts to comprehensively review all relevant information. We thank the following people who made presentations to the committee: John Dennis of the Washington office of NPS; Michael Finley, John Varley, Paul Schullery, Ann Johnson, and Wayne Brewster of YNP; Steve Torbit of the National Wildlife Federation; rancher Brian Severin; Mike Harris of Senator Conrad Burns's office; Frederic H. Wagner of Utah State University; Bob Beschta of Oregon State University; Richard Keigley, Peter Gogan, and Kim Keating of the U.S. Geological Survey; Tom Lemke and Kurt Alt of Montana Fish, Wildlife, and Parks; Clifford Montagne of Montana State University; Timothy Clark of Yale University; Rex Cates of Brigham Young University; and members of the Northern Yellowstone Cooperative Wildlife Working Group.

We hope the efforts of the committee, culminating in this report, will enhance understanding of the complexity of the ecological relationships of the ungulates of the Yellowstone northern range and will strengthen the scientific basis for effectively managing them within the context of the Greater Yellow-stone ecosystem.

All members of the committee were generous in their commitment of time through participation in the meetings, discussions, and writing of this consensus report. We appreciate the assistance provided by National Research Council staff Lee Paulson, David Policansky, Chris Elfring, Margaret Walsh, Kathy Iverson, Stephanie Parker, Mirsada Karalic-Loncarevic, Kelly Clark, Ramya Chari, and Jennifer Saunders, who supported our efforts to meet, discuss, and prepare the final report. Of particular importance to the completion of the report are the efforts of Lee Paulson, during the initial stage of committee meetings and preliminary writing; Gordon Orians, who as liaison from the Board on Environmental Studies and Toxicology, provided oversight, guidance, and critical review; and David Policansky, who undertook the difficult and final editorial task of melding and integrating the written contributions of committee members.

David R. Klein, Ph.D.
Chair, Committee on Ungulate
Management in Yellowstone
National Park

Contents

Wetland Vegetation of the Northern Range, *79*

Ecological Dynamics on Yellowstone's Northern Range

Summary

THE GREATER YELLOWSTONE ECOSYSTEM (GYE) is a large and rich temperate ecosystem that includes a variety of natural landscapes and supports diverse human activities. It provides economic, recreational, educational, and aesthetic benefits and has a growing resident human population. At the heart of the ecosystem is the 8,991-km^2 Yellowstone National Park (YNP)—declared the world's first national park in 1872, made a biosphere reserve in 1976, and added to the World Heritage List in 1978. Seven ungulate[1] species are native to the region: elk, mule deer, bison, moose, bighorn sheep, pronghorn, and white-tailed deer. All the native large predators are present: grizzly bear, black bear, coyote, mountain lion, and the reintroduced gray wolf.

YNP faces peculiar and complex management challenges. One of the most contentious issues is the management approach in place since the late 1960s called "natural regulation." Under natural regulation, ecological processes within the park generally are left to function free of direct human interventions, or, as described by the National Park Service (NPS), "natural environments evolving through natural processes minimally influenced by human actions." Concern has centered on the ecosystem of the northern range of YNP, especially about the effects of natural regulation on ungulate populations and subsequently their effects on vegetation.

---

[1]Hoofed mammals.

Ungulates that graze within YNP for much of the year often winter in the northern range in and adjacent to the park. Two-thirds of the northern winter range is within YNP; one-third is north of the park boundary on public and private lands (Figures 1-1 and 1-2 in Chapter 1). Elk and bison populations in the northern range have increased dramatically in recent years, particularly in years with mild winters, leading some scientists and members of the public to question the appropriateness of the park's natural-regulation policy. These critics believe the northern winter range is overgrazed and that woody vegetation and riparian areas are being damaged, mainly by elk. Further, they see overgrazing by elk and bison as contributing to serious erosion and stream degradation. However, other scientists and resource managers note that ungulates have influenced vegetation on the northern range for thousands of years and believe that natural density-dependent factors such as forage availability, predation, and disease are regulating population dynamics so that current conditions fall within the natural range of variability.

In recent years, the controversy over natural regulation has heightened, especially in the northern range—wintering range of Yellowstone's elk herds. In 1998, the U.S. Congress directed the NPS "to initiate a National Academy of Sciences review of all available science related to the management of ungulates and the ecological effects of ungulates on the range land of Yellowstone National Park and to provide recommendations for implementation by the Service." In response to that mandate, the National Research Council convened the Committee on Ungulate Management in Yellowstone National Park. This committee of experts was charged to review the scientific literature and other information related to ungulate populations in the Yellowstone northern range and to attempt to clarify what is known and not known about natural regulation and the ecological effects of elk and bison populations on the landscape.[2] The committee's geographic focus has been Yellowstone's northern range. The committee's evaluation addresses the issue from a scientific perspective, which deals with only part of a multifaceted problem that includes sociological, economic, aesthetic, and other important dimensions beyond the scope of this study.

---

[2]See Chapter 1 for the committee's full statement of task and a description of its methods.

NATURAL REGULATION

Natural regulation is the current NPS policy for managing ungulates and other ecosystem components within YNP. In theory, natural regulation means simply "free of direct human manipulation." The intent is to allow the biological and physical processes within the park to function without direct human intervention. A more accurate definition of natural regulation as practiced by NPS in YNP is that it attempts to minimize human impacts on the natural systems of the park. Implementing natural regulation is difficult because the NPS must accommodate the millions of visitors to YNP annually and control naturally caused fires that threaten human life and buildings. Although YNP's natural regulation policy involves little intervention within the park, ecological processes in the region are profoundly influenced by human activities outside the park.

The underlying belief that national parks should, to the maximum extent possible, harbor natural ecosystems has fostered extensive debates about how to react to ecosystem change in the parks and how to determine when such change is caused by humans. In Yellowstone, the controversy concerns whether human activities have caused ungulate populations to grow too large for the ecosystem, or whether observed changes in the ecosystem are due to natural variability in factors such as climate. If the changes are caused by humans, YNP's natural regulation policy presumably would allow intervention to mediate the effects. If they are not caused by humans, intervention would be inappropriate. The problem of differentiating human-caused from natural change is complex because no ecosystem on earth is entirely unaffected by human activity. Thus, defining "natural" is difficult. In view of the profound changes that have occurred within the GYE, such as increased development of roads and housing in areas adjacent to the park, it is no longer possible to have an ecosystem that is truly natural—that is, containing the same numbers and distributions of all the species of plants and animals that were there before European settlement, let alone before Native American populations arrived. YNP may contain many of the same species, but they can no longer respond to change as they used to by dispersal and migration.

To understand the controversy about ungulate management in YNP and evaluate management options, it is also important to understand the distinction between policy and practice. A policy is formal and is used to determine present and future decisions. A practice, which is less formal and more flexible, is the actual action of fulfilling a management concept. Natural regulation clearly is regarded as a policy by YNP resource managers. As a result, if a change in the ecosystem is natural, then management intervention is contrary to park policy, whereas if change results from an action implemented by the park or other human causes, then management is considered appropriate. If natural regulation were only YNP's practice, perhaps the debates could focus more on the actions and their outcomes and less on whether they were consistent with a policy. Also, adaptive management would be easier to pursue.

The controversy over natural regulation in Yellowstone revolves primarily around the question of whether ecological processes within the ecosystem are seriously disrupted. Because some component of the ecosystem may appear to be disrupted—e.g., effects of heavy grazing in the northern range and degradation of its riparian areas—some people criticize natural regulation. Supporters of natural regulation argue that, in the face of constantly changing biotic and abiotic environments, current conditions are within the range of natural variation and that Yellowstone is not in ecological trouble. Therefore, one of this committee's tasks was to evaluate whether conditions in the northern range ecosystem are outside the range of what might be expected based on comparisons with similar ecosystems elsewhere and historical information about the GYE. The committee also addressed whether current conditions if allowed to continue, are likely to lead to substantial and rapid change in any major ecosystem components or processes.

PERSPECTIVE: FROM PREHISTORY TO THE PRESENT

Abiotic Factors

In addition to recent human-caused changes, the ecological processes of YNP have been profoundly influenced over the long term by changes in the physical environment. Climate change has been important during the past 10,000 years. Average temperature and precipitation have changed substantially over this period and continue to fluctuate over periods as brief as a few decades. The Little Ice Age, which ended in the late 1800s, was a cooler and wetter climate than that of the 1900s, during which no substantial trend in either temperature or precipitation is evident. There have been drier and wetter times, and the past century has been characterized by somewhat less snowpack and fewer very snowy years than the previous century.

Over millennia, geological events, including mudslides, erosion, and destructive earthquakes such as the one that struck in 1959, have recurred over the centuries and continue today. In the shorter term, on the order of centuries, the influence of geological events is not usually enormous (except for cataclysmic earthquakes or eruptions), but it is noticeable and is perhaps the largest factor influencing the valley floors. The major floods of 1996 and 1997, which caused changes in the park's streams and their associated riparian communities, are examples of rapid geomorphological change.

Fires also have been part of the GYE for millennia. Major fires appear to occur every few hundred years, usually during particularly dry periods. Smaller fires occurred approximately every 20-25 years in the northern portion of YNP before the initiation of fire-control measures in the late 1800s. Natural fires were suppressed in YNP through most of the twentieth century until that policy changed in 1972. The major fires of 1988, which occurred in a very dry period, affected about 36% of YNP.

Biotic Factors

Plants

Although the northern range is primarily a shrub-steppe interspersed with Douglas-fir and lodgepole-pine forests, willows and aspen, which occupy only a small percentage of the range, have been the focus of scientific and public attention. Aspen spread clonally and typically regenerate by root sprouting with occasional episodes of seedling establishment. Thus, clones may be much older than the age of the oldest tree-sized stem. Recruitment (i.e., entry into the population) appears to have occurred through about 1920. Since 1920, however, recruitment of tree-sized aspen has been almost nonexistent. In most parts of the northern range, the sizes and aereal coverage of riparian willows and cottonwoods have decreased since the early 1900s. Most tree species that burned in the fires of 1988 successfully reseeded and regrew. Aspen vigorously reproduced by sprouting and seed germination, but browsing has prevented recruitment of tree-sized individuals. No systematic changes in abundance of other tree species have been documented, although there is evidence of expansion of coniferous forests into shrub and grassland areas.

Sagebrush and grasses have changed little at higher elevations in the northern range, locations not heavily used by elk in winter. But the cover and density of sagebrush and other shrubs has been greatly reduced at some lower elevation sites, especially areas near the northern border of YNP. Introduced timothy grass has spread widely in the northern range's Lamar Valley, and elsewhere in YNP.

Animals

Humans have used the area that is now YNP since the end of the last glaciation, about 10,000 years ago. This long history of human use has helped investigators to identify other species that were once present because humans (along with some other predators) tend to deposit animal and plant remains in concentrated areas. Based on evidence from this prehistoric period, it is clear that all the major ungulates and predators now occupying YNP, as well as many of the smaller animals, have been present for at least the past 10,000 years. However, it is impossible to estimate historical abundances of those species; all that can be stated with confidence is that they were present in large enough numbers that their remains can be found.

Reliable population estimates for elk in the northern range became available only with the initiation of aerial surveys in 1952. Historical records since the 1860s report encounters with elk, bison, grizzly bears, and other mammals, but they do not permit estimates of the densities of those mammals in and around YNP. Thus, it is not possible—and probably never will be possible—to have good estimates of the populations of elk and bison in the northern range, annually and seasonally, between about 1870 and 1920, a focal period for the controversy over aspen and other vegetation.

From the 1930s through the 1960s the NPS hunted and trapped elk to reduce their populations because park scientists during this period considered the northern range highly degraded by an excessive population of elk. By the late 1960s, when YNP adopted natural regulation, the northern range elk population had been reduced from some 10,000 animals to fewer than 5,000. But by the late 1980s, as many as 20,000 elk were on the northern range. Approximately 12,000 elk were on the northern range in 1999, with approximately 120,000 in the GYE as a whole. As of the most recent census (2000), roughly 2,500 free-ranging bison were also present in the GYE. Other ungulate populations in the GYE have also fluctuated over time; pronghorn apparently are declining, with only about 200 remaining in the late 1990s.

During various periods after the park was established in 1872, ungulates were fed in winter, hunting was restricted, and predators were killed. Predator populations (e.g., mountain lions) were reduced and wolves were extirpated. Of the major predators, the best current population estimates are for wolves, which were reintroduced to YNP in 1995. Today (2001) about 160 wolves reside in YNP, 50 of which are on the northern range. There are perhaps 300 grizzly bears, between 1,000 and 2,000 black bears, and a few hundred mountain lions in YNP and environs. Coyote populations have been reduced by wolves; for example, the coyote population in the Lamar Valley dropped from 80 in 1995 to 36 in 1998.

Although beavers in the northern range have never been accurately counted, it is clear that their populations have declined dramatically since the 1920s. The reasons for and the effects of that decline remain uncertain, although they include commercial trapping in the decade after the park's establishment. There is controversy about whether the decline in beaver populations is related to changes in aspen recruitment and changes in riparian vegetation communities.

CONCLUSIONS

Animal Populations

Factors whose influences are related to population density (called density dependent) interact with factors whose influences are not (called density independent) to regulate elk and bison populations in the northern range. There is a strong density-dependent signal in northern range elk and bison population dynamics, but their responses differ: bison tend to expand their range to areas outside YNP when their population exceeds roughly 2,500, whereas reproductive rates in elk decline when their populations exceed roughly 15,000. In addition to density-dependent factors, elk and bison populations also are affected by density-independent factors such as weather and because ungulates and their food do not always vary in a synchronous way. Thus, some ungulate populations tend to fluctuate regardless of human management intervention.

The pronghorn population has fluctuated widely during the past century and has been declining recently. Adverse factors include coyote predation and hunting on private land outside the park. Also, pronghorn may be affected by competition for food with elk, mule deer, and bison during severe winters.

Wolves also affect the population dynamics of ungulates as well as those of other predators in YNP. The nature and magnitude of the effects are not predictable at present, because the reintroduction was so recent (1995), but it is likely that wolves will reduce elk numbers. They almost certainly will cause changes in the behavior of ungulates, especially elk, including changes in areas where the elk spend time. The effect of wolves on bison is likely to be less variable and dramatic than their effect on elk, because elk are their primary prey in YNP.

Ungulates and Vegetation

Currently, in the northern range, herbivory by elk on young aspen is intense and has probably prohibited recruitment since 1920. Although there have been fluctuations in climate since 1920, none has been large enough or persistent enough to account for the failure of aspen recruitment. A plausible hypothesis—and it is no more than that at present—is that wolves, before their extirpation, affected the distribution and abundance of elk so that at least some recruitment of tree-sized aspen and tall growth of willows could occur. Recent restoration of wolves to YNP may allow evaluation of their role in aspen and willow recruitment and maintenance, but scientific information is lacking to understand the role of past development and hunting outside the park on elk behavior and migration patterns.

All tree-sized aspen in the northern range are now more than 80 years old, and in the absence of recruitment their abundance will continue to decline. Species associated with aspen will likely decline along with tree-sized stems. Elk also are reducing the size and areal coverage of willows. Defensive chemicals[3] in riparian woody vegetation may influence herbivory; however, their role in the decline in stature or loss of willows and riparian vegetation in the northern range during the past century is not known.

The architecture, size, recruitment, and coverage of sagebrush have been changed by elk, pronghorn, bison, and mule deer. The effects are more significant at lower than at higher elevations in the northern range.

---

[3]Plants defend themselves from predation in two main ways: (1) defense structures, such as thorns, and (2) toxicity and unpalatability caused by so-called "secondary chemicals." These compounds may be directly toxic or they may reduce the food value of the plant—for example, by reducing the availability of the leaf tissue protein to the animal gut.

Composition and productivity of grassland communities in the northern range have not changed much with increases in herbivory. Humans have caused some changes (e.g., the introduction of timothy grass and other exotics). Other changes might have occurred before careful inventories were taken and may not be detectable. Although conifer forests are used by ungulates, there is no evidence that ungulates affect their species composition.

Summer range does not seem to be limiting to the ungulate populations in YNP. Densities on the summer range are relatively low because the animals spread out over larger areas than in the winter range. There is little evidence for an ungulate effect on the summer range communities, with the exception of young aspen, which are heavily browsed.

The Northern Range

The condition of the northern range is different today than when Europeans first arrived in the area. This has led some people to conclude that something is "wrong" with YNP's northern range. Such conclusions reflect subjective value judgments in addition to objective observations. For example, some people compare the northern range unfavorably with nearby ranches, but that reflects a mixing of values. Ranching seeks high production for human uses, but YNP seeks to preserve a natural environment and the species and ecological processes within it. The committee recognizes that such value judgments influence debates about YNP but has focused this report on scientific information and conclusions.

The committee judges that the changes in the northern range are the result of the number of ungulates in the area combined with biophysical factors such as climatic variability, but current methods do not allow us to separate the relative contributions of each of these effects. However, the committee concludes, based on the best available evidence, that no major ecosystem component is likely to be eliminated in the near or intermediate term. Further, although we recognize that the current balance between ungulates and vegetation does not satisfy everyone—there are fewer aspen and willows than in some similar ecosystems elsewhere—the committee concludes that the northern range is not on the verge of crossing some ecological threshold beyond which conditions might be irreversible. The same is true of the region's sagebrush ecosystems, despite reductions in the number and size of plants at some lower elevations.

Natural Regulation

The conclusions in this report should not be interpreted either as a criticism or as a vindication of YNP's natural regulation policy other than to say that it has not been associated with ecological disaster. "Natural" cannot be unambiguously or objectively defined. In addition, human activities adjacent to YNP have large effects on the animals present at least seasonally within the park. The animals do not have free access to the adjacent areas that formerly were available to them as migration corridors and winter range. For these reasons, true natural regulation in YNP—that is, really letting nature take its course with no human intervention—is not possible.

YNP's practice of intervening as little as possible in the ecology of ungulates within YNP will likely allow the persistence of the northern range ecosystem and its major components as long as there is no large change in climate. If the NPS decided that it needed to intervene to protect species like aspen and the species that depend on tree-sized aspen stems, localized interventions would be prudent. For example, if YNP decided to maintain tree-sized aspen in the park, putting exclosures around some stands would be less potentially disruptive than eliminating ungulates or reducing their numbers. The most effective way to reduce ungulate numbers in YNP would be to shoot them (unless wolves have a larger effect than currently expected). Earlier shooting of elk in YNP provoked strong public protest. Without strong scientific justification for doing so, which the committee cannot provide, future shooting of elk in YNP would provoke strong public protest again, and its benefit would be unclear at best.

We emphasize again that large ecosystems in general and YNP's northern range in particular are dynamic. Ecosystems change in unpredictable ways. The recent addition of wolves, which has restored an important component of this ecosystem, adds to the dynamism and uncertainty, especially in the short term. Whether viewed as an experiment or not, the near future promises to be most instructive about how elk and other ungulates interact with a complete community of predators.

RECOMMENDATIONS

Given the complexities involved in managing Yellowstone's dynamic ecosystems, there is a continuing need for rigorous research and public education.

The committee offers the following recommendations designed to enhance understanding of the key processes that affect Yellowstone's ungulate populations, vegetation, and ecological processes.

Park Management and Interpretation

• To the degree possible, all management at YNP should be done as adaptive management. This means that actions should be designed to maximize their ability to generate useful, scientifically defensible information, including quantitative models, and that the results of actions must be adequately monitored and interpreted to provide information about their consequences to guide subsequent actions.

• There is insufficient scientific knowledge available to enable objective comparison of different management approaches and understanding of the consequences of management choices. Thus, long-term scientific investigations and experiments to provide solid scientific evidence for evaluating management options are needed.

• The NPS educational and outreach program can play an important role in fostering public understanding of the complex and dynamic nature of ungulate ecology in the GYE, which is an essential adjunct to effective management of the northern Yellowstone ungulates. In this regard, we encourage the NPS to increase their focus on entire ecosystem relationships, processes, and dynamics of the GYE, especially emphasizing the importance of primary production (conversion of sunlight to stored carbon by plants) and trophic-level (i.e., hierarchical levels in the food web) relationships.

Vegetation

• A rigorous study focusing on current aspen populations throughout the GYE should be undertaken to quantify the relative importance of the factors known or hypothesized to influence aspen stand structure. It should include the use of an increased number of large exclosures with a long-term commitment to monitoring the effects of restricting herbivory by ungulates. The study sites could be discussed in the NPS ecosystem interpretive program.

• The most important driving variables that affect the modified riparian ecosystems in these areas today, especially the relationship between herbivory and groundwater availability, need to be carefully examined. This should include an understanding of fluvial processes, surface and groundwater hydrology, and biotic processes.

•   Research and monitoring should continue on northern-range sagebrush–grassland communities.

•   Research to determine whether it is possible to differentiate ungulate use of tall and short willows on the basis of both the food-deprivation levels of the ungulates (i.e., winter starvation) and levels of defensive chemicals in the plants is needed.

Animal Populations

•   The behavioral adaptations of elk and other ungulates as well as changes in their patterns of habitat use as a consequence of the presence of the wolf as a large predator newly restored to the system should be closely monitored as a basis for understanding the dynamic changes that are taking place within the system.

•   The changes taking place in the interactions among the large predators of YNP and their effects on the trophic dynamics of the ecosystem should be closely monitored as the reintroduced wolves become an established component of the system.

•   Thorough study of current and likely future trajectories of the pronghorn population and the role of human impacts on this population, including disturbance by visitors and the Stevens Creek bison facility, is needed. The study should evaluate the likely consequences of a full range of potential management options from doing nothing to actively controlling predators and providing winter feed.

•   Periodic surveillance for pathogens (including brucellosis) in wild ruminants in the northern range should be continued, and a more thorough understanding of population-level threshold dynamics gained. Samples could routinely be obtained from animals immobilized for research, found dead, or killed by hunters.

Biodiversity

•   A periodic and comprehensive biodiversity assessment every 10 to 15 years is needed on the northern range to evaluate potential direct and indirect impacts of ungulate grazing, for both terrestrial and aquatic environments. Indicator species that reflect habitat change should be identified. Those species should be monitored intensively between comprehensive assessments.

Human Influence

•   A comprehensive research effort is needed to assess the influence of seasonal densities, distribution, movements, and activities of people within YNP and adjacent areas on wildlife species, their habitat use patterns, behavior, foraging efficiency, effects on vegetation, and other aspects of their ecosystem relationships.

•   The effects of changing land-use patterns in the landscape surrounding Yellowstone must be understood with regard to its expected influence on the park's biota and natural processes, such as fire.

EPILOGUE

GYE is dynamic, and change is a normal part of the system as far back as we have records or can determine from physical evidence. Based on that record of change, it is certain that sooner or later the environment of GYE will change in ways that cause the loss of some species and changes in community structure. If human-induced changes are taken into account, both within GYE and globally, that circumstance is likely to be sooner than would otherwise occur.

Although dramatic ecological change does not appear to be imminent, it is not too soon for the managers of YNP and others to start thinking about how to deal with potential changes. Before humans modified the landscape of the GYE—limiting access to much of lower elevation wintering areas and interrupting migration routes—animals could respond to environmental changes by moving to alternative locations. Over a longer time frame, plants could adapt as well, although to a lesser degree, especially in places with significant topographic relief. But many options that organisms formerly had for dealing with environmental changes have been foreclosed because of human development in the region. Human-induced climate change is expected to be yet another long-term influence on the ecosystem. A future challenge for the GYE area and other wildlands will be reconciling the laudable goals of preserving ecosystem processes with human interests and influences. That reconciliation will require resolving conflicting policy goals, bolstering incomplete scientific information, and overcoming management challenges. Doing so will require all the vision, intellectual capacity, financial resources, and goodwill that can be brought to bear on them.

1

Introduction

BACKGROUND AND CHARGE TO THE COMMITTEE

THE GREATER YELLOWSTONE ECOSYSTEM (GYE) is a large and rich temperate ecosystem (Figure 1-1). It includes six national forests, Yellowstone National Park (YNP), Grand Teton National Park, and two national wildlife refuges. It contains the largest functional geothermal basin in the world. More than 1,700 plant species have been identified in the GYE; 80% of the area is forested. Ten species of fish, 24 of amphibians and reptiles, more than 300 of birds, and 70 of mammals are present in the GYE in addition to thousands of invertebrate species. This rich and varied ecosystem supports diverse human activities; provides economic, recreational, educational, and aesthetic benefits; and has a growing resident human population. At the heart of the ecosystem is YNP—established as the world's first national park in 1872, made a biosphere reserve in 1976, and added to the World Heritage List in 1978.

YNP, which encompasses 899,139 ha (8,991 km^2), faces peculiar and complex management challenges. In the 128 years since YNP was established, a variety of management policies and strategies have been used to fulfill the park's mission and its relationship to the American public's desire for expanded tourism and recreational opportunities. The concurrent changes in societal values and attitudes toward the natural environment have complicated management of YNP over this time. Adding to the management challenges is the knowledge that management approaches implemented in YNP, the

15

FIGURE 1-1 The Greater Yellowstone Ecosystem. Source: Clark et al. 1999. Reprinted with permission; copyright 1999, Yale University Press.

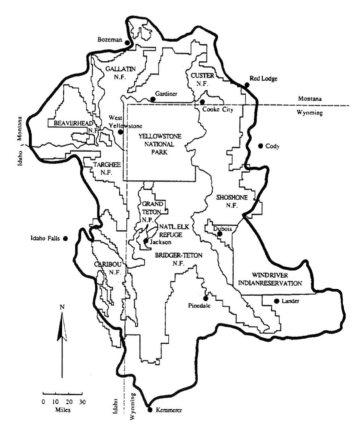

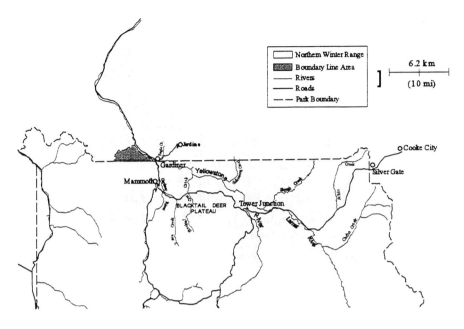

FIGURE 1-2 Yellowstone's northern winter range. Source: YNP 1997.

nation's premier park, influence management in other parks in this country, and throughout the world, as well as affect the economic bases of neighboring states, where wildlife viewing, recreation, and hunting are important. One of the most contentious approaches, applied since the late 1960s, is the policy of "natural regulation." Concern has centered around the effects natural regulation might have on ecosystem processes, particularly in the northern winter range of ungulates of the GYE (Figure 1-2). Under natural regulation, ecological processes and physical influences—such as primary production, foraging, competition, weather, predation, and animal behavior—determine or limit population dynamics, rather than hunting and other human interventions.

Seven ungulate[1] species are native to YNP: elk (*Cervus elaphus*), mule deer (*Odocoileus hemionus*), bison (*Bison bison*), moose (*Alces alces*), bighorn sheep (*Ovis canadensis*), pronghorn (*Antilocapra americana*), and white-tailed deer (*Odocoileus virginianus*). Mountain goats (*Oreamnos americanus*) are not native to the park, but their numbers are increasing from

[1]Hoofed mammals.

introductions made outside YNP, particularly in the Absaroka, Gallatin, and Crazy Mountain ranges of Montana. All native large predators are present, including the grizzly bear (*Ursus arctos*), black bear (*Ursus americanus*), coyote (*Canis latrans*), mountain lion (*Felis concolor*), and the reintroduced gray wolf (*Canis lupus*).

Ungulates that graze within YNP for much of the year often winter on the northern range in and adjacent to YNP. The northern range, a mixture of grassland and forest approximately 153,000 ha in area, encompasses lands along the Yellowstone River and Lamar River basins lower than 2,255 m from the junction of Calfee Creek and the Lamar River in the east to the area around Dome Mountain and Daly Lake in the west. Two-thirds of the northern winter range is within YNP; one-third is north of the park boundary on public and private lands (Houston 1982, Clark et al. 1999) (Figure 1-2). The pre-European winter range included the lowlands north of the park now devoted to agriculture, ranching, and rural residences.

Elk and bison populations have increased markedly during the past century, particularly in years with mild winters, leading some scientists to question the appropriateness of the park's natural-regulation policy. Claims have been made that the northern winter range is overgrazed and that woody vegetation and riparian areas are being destroyed. Because they are the most abundant ungulate species, elk have been held primarily responsible for these effects. Elk commonly browse woody vegetation, such as aspen, cottonwood, and willow, especially in winter. The claim is that they have eliminated much of the riparian-zone woody vegetation and are preventing generation of aspen and cottonwood stands, thereby reducing food sources for other species, such as beavers, moose, deer, and grizzly bears. Stream degradation and serious erosion have also been attributed to overgrazing by elk and bison. The spread of diseases, especially brucellosis, among dense populations also is a concern. However, other scientists claim that ungulates have influenced vegetation on the northern range for thousands of years when natural density-dependent factors such as forage availability, predation, and disease regulated population dynamics and that current effects fall within the natural range of variability.

Because the controversy over natural regulation has heightened recently, in the 1998 appropriations to the U.S. Department of the Interior, Congress (the House Appropriations Committee report) stated: "A number of scientists question the natural regulation management program conducted by Yellowstone National Park as it relates to bison and elk, while others defend the approach. The Committee wishes to resolve the issue of population dynamics of the northern elk herd as well as the bison herd. The Committee thus directs the [National Park] Service to initiate a National Academy of Sciences (Board on Environmental Studies and Toxicology) review of all available science related to the management of ungulates and the ecological effects of ungulates on the range land of Yellowstone National Park and to provide recommendations for implementation by the Service" (HR Report 105-163; appropriations in 105-83).

In response to that mandate, the National Research Council convened the Committee on Ungulate Management in Yellowstone National Park. The committee is composed of experts with backgrounds in ungulate ecology, wildlife biology, animal/veterinary science, animal population modeling, grassland ecology, riparian ecology, climatology, hydrology and geomorphology, landscape ecology, and soil science. These experts were charged to review the scientific literature and other information related to ungulate populations on the Yellowstone northern range, particularly as they relate to natural regulation and the ecological effects of elk and bison populations on the landscape. Much of YNP is not noticeably affected by current management practices. The northern range, however, where Yellowstone's elk herds spend the winter, is the origin of much controversy, and the geographic focus of this study. The following specific scientific questions were addressed within the context of the park's goals:

- What are the current population dynamics of ungulates on the northern range of the GYE?
- To what extent do density-dependent and density-independent factors determine average densities and fluctuations in populations of YNP ungulates?
- What are the consequences of continuing the current natural regulation practices (e.g., on range condition, habitat for other species, risk of disease transmission)?
- How do current ungulate population dynamics and range conditions compare with historical status and trends in those processes?
- How do current ungulate population dynamics in the GYE compare with other North American grasslands and savanna ecosystems that still have native large predators?
- What are the implications and limitations for other natural regulation practices applied to other biota?
- What gaps and deficiencies in scientific knowledge should future research attempt to address?

During its analyses, the committee identified other topics relevant to its charge and important to evaluating the consequences of current ungulate management strategies in YNP.

The committee met four times over the course of its deliberations. Meetings were held in Gardiner, Montana, and in Mammoth Hot Springs, Wyoming, to permit committee members to see the northern range in summer and winter and to obtain input from federal and state agency personnel and members of the public who are intimately familiar with the issues. The committee is aware of the deeply held convictions of many different parties and the difficulties faced by land and animal managers in the GYE. This consensus report presents the committee's scientific evaluations of the issues it was asked to address. The committee recognizes that these evaluations are only part of a multifaceted problem that includes sociological, economic, and other important dimensions that it was not asked to evaluate.

NATURAL REGULATION

Natural regulation is the current National Park Service (NPS) policy for management of ungulates and other ecosystem components within YNP. As currently practiced, natural regulation means simply "free of direct human manipulation." The intent is to allow the biological and physical processes within the park to function without human influence. Implementation of the policy includes letting wildfires burn freely and prohibiting hunting within the park. However, NPS has other policy mandates within YNP that might not be consistent with natural regulation. For example, the National Park Organic Act of 1916 prescribes that "[t]he fundamental purposes of said parks is to conserve the scenery and the natural and historic objects and the wildlife therein and to provide for the enjoyment of the same in such a manner and by such means as will leave them unimpaired for the enjoyment of future generations." It is of course not possible to construct and maintain a road system within the park, with associated public facilities and services, "to provide for the enjoyment" of the three million visitors who currently visit the park annually, without affecting the scenery, vegetation, and animal life.

A more realistic definition of natural regulation as practiced by NPS in YNP is that it attempts to minimize human effects on the natural systems of the park. The National Park Service 1988 Management Policies describes the management strategy as "natural environments evolving through natural processes minimally influenced by human actions." Under this policy, some interventions have been allowed. For example, to restore natural predation processes, wolves were reintroduced. To gain a better understanding of ungulate ecology in the park, there has been some manipulation of animals and vegetation through live capture, the equipping of animals with radio transmitters, and the use of fenced enclosures to assess plant growth in the absence of grazing and browsing. Research is also undertaken to satisfy NPS mandates to better facilitate management of visitors to the park, to minimize their impact on wildlife and vegetation, and to enhance public enjoyment by providing interpretive information about the ecology of the park. Where these mandates conflict with preservation of natural environments and processes, conservation is considered the park's primary responsibility (NPS 2001).

NPS also influences the behavior and movement patterns of ungulates and their predators through control and routing of park visitors in both summer and winter. YNP biologists and managers participate in the Northern Yellowstone Cooperative Wildlife Working Group, which includes NPS; Montana Fish, Wildlife and Parks; Gallatin National Forest; and the Biological Resources Division of the U.S. Geological Survey. They also provide advice and make recommendations in the development of management policy for YNP ungulates when they are outside of the park, where human interventions are more pervasive.

Although YNP's natural regulation policy attempts to minimize human intervention within the park, this does not characterize the policies that affect ungulates when they are outside the park. In other words, YNP is an ecological island whose processes are influenced by human activities in the surrounding area. These activities, which strongly influence YNP wildlife, include agriculture, ranching, and hunting. Thus, even if there were no human intervention within YNP, ecological processes there would be profoundly influenced by human activities elsewhere. In this sense, management can, at most, be only partly natural.

What Is Natural?

Climate displays a great deal of variability, even over the ecologically short period of 100 to 200 years. Other ecological processes also fluctuate with time but fluctuations may fall outside natural variability if driving factors create extraordinary responses. The limits of variability of an ecosystem process or component can be known only if those attributes have been examined with appropriate tools using long-term analyses. This has not been the case for YNP. How, then, can a reasonable standard be found to indicate what range of conditions on YNP's northern range is natural? The historical sequence of interacting biotic, environmental, and anthropogenic factors is unrepeatable, which makes it difficult to identify "natural baseline conditions" (Anderson 1991, Patten 1991). It follows, therefore, that if the natural regulation policy assumes that all processes under the policy are functioning in a natural fashion, any outcome of the policy would also have to be termed natural. However, our lack of understanding about previous conditions and current dynamics makes it difficult to reject either the hypothesis that ungulate populations will be naturally regulated without causing long-term damage to the GYE or the notion that naturally regulated ungulate populations will cause long-term damage to the GYE. Outcomes may indeed be natural, but their desirability is a question of values beyond this committee's scientifically circumscribed task.

Some Implications of a Natural Regulation Policy

The national park system comprises many diverse parks, ranging from vast expanses of near-wilderness, like YNP, to small urban gardens, such as Kenilworth Aquatic Gardens in Washington, D.C. No single management policy is appropriate for this diverse array of parks.

A natural-regulation policy is motivated in part by the belief that national parks should contain natural ecosystems. However, even the largest of them is only a fragment of the predevelopment ecosystem, and none has been unaffected by human activity. So the "natural" in "natural regulation" can never be absolute, and debates about what interventions are appropriate will inevitably accompany all management decisions. Pritchard (1999) described the issue well: "A rock-solid definition of what is natural will elude us, because that question is wrapped up in our cultural attitudes about our place in nature. In an age when science recognizes the role of change and disturbance in the landscape, discussion will shift from 'how do we restore nature's balance?' to consideration of how much change we are willing to allow in our parks. Given the shrinking size of naturally forested areas and continually growing human disturbances in the region, however, landscape changes may bump into other conservation goals, namely the preservation of threatened species."

The belief that national parks should harbor natural ecosystems has fostered extensive debates about whether observed changes in the parks are caused by humans. If they are, natural management would require intervention to reduce, if possible, human-caused changes. If they are not, intervention would be inappropriate. The problem is aggravated by changes in the perception of the role of humans in nature, from acceptance that Native Americans were part of the GYE in the past to today's viewpoint in our highly urbanized society that humans exist outside of nature. The problem, as described by Pritchard, is to distinguish between natural and human-caused changes. Disagreements about this issue are at the heart of controversies over management of YNP. For example, critics of YNP maintain that human influences on elk populations, and on their predators, have led to overgrazing. Thus, overgrazing represents a human impairment and should be managed. In contrast, YNP maintains that changed climate is a major driver and that the ecological changes are natural. As Schullery (1997) put it, "[W]e have imagined ourselves wise enough to control [Yellowstone National Park] and have rushed to judge what is wrong with it. And every time we looked hard enough, we discovered that there was more wrong with our judgment than with Yellowstone."

Although such debates in part may be stimulated by the value-laden word "natural regulation," describing management strategies by different words will not cause the debates to disappear. Controversy will continue, because, fundamentally, the debates are not about the words but rather about the roles people want the parks to play.

Policy Versus Practice

The distinction between policy and practice is subtle and ill defined; however, an understanding of the distinction as applied in YNP is important for understanding the controversy about ungulate management in YNP and for evaluating management options. A policy is "a definite course or method of action selected from among alternatives and in light of given conditions to guide and determine present and future decisions," whereas a practice is much less formal or rigid, being "actual performance or application, a repeated or customary action, [or] the usual way of doing something" (Merriam-Webster 1993).

Natural regulation as a policy for YNP (Schullery 1997, YNP 1997, Pritchard 1999) implies that if a change in the ecosystem is natural, then management intervention is inappropriate. On the other hand, if a change is due to a management policy implemented by the park, then there is justification for repair or restoration of that change or impairment. This has led to distracting and ultimately unproductive arguments about whether high elk populations and low aspen recruitment in the northern range are the effects of natural changes (e.g., climate) or human actions (e.g., removal of predators). Under a policy of natural regulation, such an argument is important, because the appropriateness of the park's management actions (or lack of them) depends on whether they are consistent with the policy. If natural regulation were only a practice, the debates could focus on the actions and their outcomes. In addition, adaptive management might be easier to undertake.

The committee believes that a better way to approach these issues is to focus on objectively measured processes, numbers, and events and how intervention might alter them rather than to debate what is or is not natural. Thus, this committee has tried to assess whether current ecosystem conditions are outside the range of what might be expected in similar ecosystems elsewhere or might have been expected in this ecosystem over the past few millennia. The committee has also tried to assess whether current conditions are likely to lead to a substantial and rapid change in any major ecosystem component or process.

HISTORICAL CONTEXT

The current policy for management of YNP has been influenced by the history of the region, the park's establishment, previous park management, and the public's goals for the park (Schullery 1997, Pritchard 1999). One ongoing debate focuses on whether the "natural" in natural regulation includes humans as part of nature. Most writing on the subject relative to North America has included aboriginal people and their activities before contact with Europeans (Callicott 1982). But historical and archeological evidence suggests that Native American cultures and their effects on the environment have been in a dynamic state of flux marked by periods of change and intervals of stasis since the first humans arrived from Asia (Guthrie 1971). In the Great Basin and adjacent Rocky Mountain region, human cultures were based on hunting and gathering, but during the century before direct contact with Europeans, major changes in aboriginal peoples' distribution, numbers, and cultural relationships to the environment followed acquisition of the horse (an indirect product of European contact). Thus, whether we view humans as part of nature, we still must acknowledge that humans have, since their arrival in the area, been agents of change.

In addition to human-caused changes, the ecology of YNP has been profoundly influenced by changes in the physical environment. Of these, climate change has been the most important driving factor during the past 10,000 years. Climate models developed by scientists project major climate changes for the future, some of them caused by human activity (NRC 2001). Climate changes complicate management strategies because they make it especially difficult to predict the likely consequences of any human interventions as well as the consequences of not intervening.

Biodiversity Context

Yellowstone's wildlife, including its ungulates, is part of a diverse biota. YNP is estimated to have 1,700 species of native vascular plants, 170 exotic plant species, 186 species of lichens, 59 mammal species, 311 species of birds, 18 fish species, 6 species of reptile, 4 amphibian species, and an unknown number of organisms associated with hot springs, to say nothing of insects, invertebrates, fungi, and bacteria (YNP 2001). The GYE contains even more species. In addition, YNP is home to several threatened (grizzly bear, Canada lynx, and bald eagle) and endangered (peregrine falcon, timber wolf, and whooping crane) species.

Biodiversity is a dynamic ecosystem characteristic that may fluctuate naturally or in response to human activity (Noss and Cooperrider 1994). Altered community structures, and processes including those resulting from human activities and management practices (Weins and Rotenberry 1981, Rothstein 1994, Hansen and Rotella 1999), may change species composition and prevalence (Hansen and Urban 1992), but how they have done so in the GYE is poorly understood. Changes caused by changes in population dynamics of ungulates may affect not only dominant landscape features, such as aspen or sagebrush, but also vast numbers of associated organisms.

A Brief History of Park Management

At least 28 federal, state, and local entities manage various activities in the GYE, sometimes with conflicting goals. Within YNP, management policies generally have been based on the prevailing understanding of wildlife biology and ecology. When the park was established in 1872, ungulates were fed and given protection from hunting and predators because they were considered the species visitors desired to view. Most predators were reduced, and wolves were extirpated from the area. Wolves were restored in 1995; about 160 are present in the GYE, 50 of which are on the northern range. From the 1930s through the late 1960s, NPS shot and trapped elk to reduce the population to what was considered a sustainable level for the northern range (Houston 1982). By the late 1960s, the elk population on the northern range had been reduced from some 10,000 animals to fewer than 5,000. Unfavorable reactions to NPS control measures from a variety of public interests led to Senate hearings. YNP then adopted natural regulation practices because they had no other management alternative (Pritchard 1999). The most recent northern range elk count observed 13,400 individuals (T. Lemke, Montana Fish, Wildlife and Parks, personal communication, June 13, 2001). Some 120,000 elk are present in the GYE. Roughly 2,500 free-ranging bison also are present in YNP.

Aldo Leopold (1933), in his classic *Game Management,* which laid the foundation for modern wildlife management, acknowledged that wildlife management was essentially management of human behavior toward the land that harbored wildlife and toward the wildlife through harvest by hunting and trapping. To quote Leopold, ". . . game management produces a crop by controlling the environmental factors which hold down natural increase, or productivity, of the seed stock" (Leopold 1933). Management of wildlife within YNP, or more correctly management of human behavior that affects wildlife within YNP, has been driven by the view that national parks exist for public enjoyment and appreciation. Thus, wildlife within YNP has been managed to achieve certain objectives for human interactions with wildlife, although recent management philosophy tends toward ecosystem management. To enhance human contact with wildlife, roads, trails, campgrounds, and other park facilities have been constructed. Those actions have altered habitats upon which wildlife depends and have influenced the movement, location, and concentration of people, which further alters the distribution of wildlife.

YNP was established in 1872, more than 60 years before Leopold provided the conceptual basis for modern wildlife management. Efforts to manage wildlife in Yellowstone in the past and present must therefore be viewed within the context of those times (Table 1-1). Market hunting of ungulates and killing of carnivores were common in the area before and continued after the park was established. In 1883, public hunting within the park was prohibited, but no effective enforcement occurred until the U.S. Calvary was assigned in

TABLE 1-1 Chronology of Significant Events in Yellowstone National Park's History

Year/Period	Event
1860s	Period of extensive fires (Romme and Despain 1989)
Before 1870	Evidence of substantial elk use of the northern range, at least in summer (Houston 1982)
1872	Yellowstone National Park declared first national park (organic act)
1869-1883	Extensive market hunting; ungulates and carnivores greatly reduced (YNP1997)
1883	Public hunting within YNP prohibited (YNP 1997)
1886	U.S. Calvary assigned to protect the park; beginning of effective control of hunting (Houston 1982, YNP 1997)
1894	Lacey Act enacted, prohibiting all hunting and killing of wildlife except dangerous animals (Schullery 1997)
1900-1935	Intensive control of predators; wolves extirpated (YNP 1997)
1902	Few bison remained in YNP; population supplemented from domestic bison herds (Meagher 1973)
1917	Brucellosis detected in YNP bison (Mohler 1917)
1918	U.S. National Park Service assumed control of YNP (YNP 1997)
1920s	"Too many elk in park"; active management to control population. Increasing concern about overgrazing (YNP 1997)
1920s	Reported decline in white-tailed deer population from about 100 to few or none (Skinner 1929)
1920-1960	Commercial definition of over-grazing applied to YNP; intensive population control of elk and bison (YNP 1997)
1923-1929	Elk removed primarily by hunting outside park; probably 10-15,000 elk on northern range (Houston 1982)
1930s-present	Very little recruitment of aspen on the northern range
1960s	Period of most intensive elk population control and population reduction (YNP 1997)
1963	Leopold report published (Leopold et al. 1963)
1968	YNP adopts policy of "natural regulation"; intensive regulation of elk and bison ends (Cole 1971)
1969-1981	Period of rapid increase in elk population from ~4,000 to ~16,000 (Houston 1982)
1969-1995	Bison population expands from ~400 to ~4,000 (YNP 1997)
1986	Congress funds study of "overgrazing" in the northern range
1988	Extensive fires in YNP
1988-1989	Severe winter reduces elk and bison populations
1995	Reintroduction of wolves
1996-1997	Severe winter; slaughter of >1,000 bison as they left YNP

1886 to protect the park. Public attitudes toward wildlife at the time were mixed (Magoc 1999). With the rapid expansion of domestic livestock onto western range lands, many viewed wild ungulates as competitors for forage. Others appreciated them for their meat and hides. The decline of wild ungulates in the Yellowstone area by the end of the nineteenth century resulted largely from market hunting (Krech 1999). Construction of the Northern Pacific Railroad into the region north of the park in 1883 expanded access to markets for meat and hides. Concern about the decline of wild ungulates developed among residents in the area as well as in eastern states where an embryonic national conservation movement was emerging. At that time large carnivores were viewed as undesirable by most of the public, because they killed both domestic livestock and wild ungulates and were feared to pose a threat to humans. As a reflection of this public sentiment, although other wildlife species were protected within the park, control of predators by park managers began in the nineteenth century and became intensive between 1900 and 1935, resulting in the extirpation of wolves and probably mountain lions.

Few bison remained in Yellowstone by the end of the nineteenth century, and the population was supplemented from domestic herds in 1902 (Meagher 1973). The Yellowstone bison herd had become the only free-ranging herd in the United States, and thus, the YNP played a key role in restoring bison in the wild and removing the threat of their extinction that existed by the late 1800s. Intensive management of bison within the park was the practice until the NPS announced in 1968 a policy of natural regulation of wildlife populations. Management included an early period of annual roundups with selective slaughter, maintaining an irrigated hay field in the Lamar Valley for winter feeding, capturing and shipping animals outside of the park, and, from the 1920s into the late 1960s, harvests to prevent overgrazing on the winter range (Pritchard 1999). Concern about the spread of brucellosis, carried by bison, to domestic livestock outside of the park led to the slaughter of more than 1,000 bison as they left the park during the extreme winter of 1996-1997 (Peacock 1997).

In the twentieth century, although not as intensively managed as bison, elk in Yellowstone were periodically fed in winter and were subjected to population reductions through shooting, corralling, and translocation until the winter of 1968-1969, a year after the adoption of natural regulation (Pritchard 1999). Hunting of Yellowstone elk that leave the park in winter has continued throughout the history of the park. Other ungulates present in the northern portions of YNP—mule and white-tailed deer, moose, bighorn sheep, and mountain goats—have not been the focus of restoration efforts or population reduction efforts in response to concerns about overgrazing.

In the early history of the park, the public and park managers thought that wolves, mountain lions, and coyotes threatened the well-being of the park ungulates. The ungulates, because of their attractiveness to park visitors and the opportunities that existed to observe them, were considered a primary justification for the park's existence and thus were the focus of management efforts. Interestingly, bears, which were considered a threat to human interests outside of the park, were appreciated by park visitors, who were allowed to view them at garbage feeding sites until the practice was discontinued in 1943 (Pritchard 1999).

By the middle of the twentieth century, there was a broadening appreciation for wildlife among the public along with an increased understanding of the interrelationships of wildlife in nature. This increased interest in wildlife led to recognition of the importance of natural environments for wildlife as well as their importance in providing ecologically rich environments for Americans to appreciate nature through leisure travel and other outdoor activities.

Changes in management of wildlife in YNP since its establishment reflect the changes in attitudes toward wildlife that have taken place among the American public. Several research efforts aimed at increasing understanding of the ecology of ungulates and carnivores within the park were initiated in the post-World War II years, although many studies occurred earlier and are cited in several sections of this report. Two studies worthy of note were by the Muries. In the 1930s Adolph Murie (1940) completed a detailed study of the diets of coyotes in the park, and his brother, Olaus Murie (1951), studied elk foraging ecology in the Jackson Hole area south of the park. Research findings by the Muries and other researchers who followed them, although preliminary and often controversial (Leopold et al. 1963, Schullery 1997), disclosed some of the complexity of the ecological relationships of bears and coyotes as well as of the ungulates in the park. Large predators had finally gained appreciation in the eyes of both the public and the park managers. They were protected with the intent to restore them as valued components of the park wildlife. However, the NPS did little to encourage continued research toward understanding the complexity of predator-prey relationships or the details of the functioning of park ecosystems (Pritchard 1999). As a consequence of carnivore protection in the park, NPS came under increased pressure from ranchers outside of the park who were losing livestock to coyotes, bears, and mountain lions that were leaving the park as their populations increased.

In 1967, NPS Director Hartzog, along with Secretary of the Interior Udall and Senator McGee of Wyoming made a decision to stop killing elk in YNP. Hartzog announced that the most desirable means of controlling the elk numbers was through public hunting outside the park. According to Sellars (1997), this policy decision to use "natural regulation" for management of wildlife "came not as the result of scientific findings, but because of political pressure." Whereas Cole's account (1969) represents natural regulation as simply a research hypothesis. This was essentially a "hands off" policy that abolished direct manipulation of wildlife or their habitats except to repair or reduce damage caused by human activities (Schullery 1997), although early elk management practices under this policy still allowed trapping and transport of elk (Sellars 1997). Recently, some intervention has been considered appropriate, including reintroducing wolves to the park to reestablish their role as a major predator of ungulates. Criticism of natural regulation policy and its perceived consequences resulted in the establishment of this committee and this report.

CLIMATE VERSUS ELK AS CAUSALITY OF CHANGES IN ASPEN AND WILLOW

The controversy over the status of aspen and willow within the northern range (often considered indicators of ecosystem condition) and natural regulation practice often revolves around conflicting statements about what factors most strongly influence the conditions currently found in the northern range. Supporters of natural regulation policy believe that current range conditions are determined primarily by biophysical factors, climate in particular, and that ungulate populations are having a small effect on range condition. Others believe that ungulate populations play a central role in determining range condition and that those populations are now so large that damage to vegetation on Yellowstone's northern range has occurred. As an example, the impact of elk on streamside willows is a central issue in the controversy. However, willow communities are inextricably linked to climate, geology, surrounding vegetation, and fire history as well as to shorter-term and local impacts from ungulates and beavers, which makes the issue hard to resolve.

Several people who made presentations to the committee compared the condition of the northern range unfavorably with that on nearby ranches. Such comparison, however, reflects a confusion of values. Ranching seeks to maximize production for human use, whereas YNP seeks to preserve natural ecological processes. For example, within YNP, aspen is an important landscape element—it is an aesthetically pleasing tree that grows in groves on the hillsides and shows spectacular fall colors. It also is important as habitat for a diversity of bird species (Mueggler 1988, Pojar 1995). Willow, by comparison, is a rather plain plant that most lay visitors to YNP do not notice or might consider an ordinary "bush" if they did. Nevertheless, willow communities may prevent stream-bank erosion and subsequent changes in stream-channel morphology and are a food source for browsers and habitat for many other riparian species. Critics of the park's natural regulation policy particularly emphasize the adverse effects of elk on streamside willows, which they maintain are excessive and are leading to erosion of northern range streams (Kay 1990, Chadde and Kay 1991).

ORGANIZATION OF THIS REPORT

To present a perspective for understanding current conditions in YNP, Chapter 2 describes historical conditions in the GYE, including climate, geological, and landscape conditions. Chapters 3 and 4 give an ecological context in which park management strategies are conducted, including a review of vegetation, possible driving factors, and processes related to ungulate population dynamics. Chapter 5 is a synthesis chapter that reviews overarching questions related to the problem at hand, and it presents the committee's conclusions and recommendations.

2

Historical Perspective: Yellowstone's Changing Environment

EVALUATING THE CONSEQUENCES of a natural regulation policy requires that the long-term dynamics of the Yellowstone landscape be understood in terms of variation in climate, disturbance regimes, vegetation patterns, animal populations, and human occupancy or use of the landscape. Yellowstone is a dynamic landscape, and we cannot determine whether management actions have forced components of the system beyond their historical range of variability unless we place recent dynamics in a longer time frame. Knowledge of prehistoric and historical environments is essential for creating a context for this evaluation. This chapter points out how dynamic the landscape of the GYE has been over time.

SCALES OF CHANGE

On a geological time scale, earth's history is one of continuous change driven primarily by plate tectonics, with periodic extraplanetary influences, such as asteroid impacts and solar cycles. On this scale, the earth is fractured; subducted; uplifted; built up by volcanism and sedimentation; and worn away by ice, water, wind, heat, and gravity. Yellowstone National Park (YNP) is centered on a geological mantle plume, or hotspot, where the molten magma comes close to the surface (Anders et al. 1989, Pierce and Morgan 1992, Pritchard 1999). The North American plate is thought to have moved to the southwest, leaving a series of volcanic traces from southwestern Idaho along the Snake River to the currently active Yellowstone region. Geomorphological changes have not stopped, for uplifting is continuing (Reilinger 1985). YNP is characterized by boiling mud pots, thermal pools, geysers, a volcanic caldera, ancient lava flows, and eroding rivers and waterfalls, which were the original reason for its protection as a park—wildlife and other life forms were an afterthought. Geomorphological changes may proceed gradually or may occur abruptly. Yellowstone witnessed a major earthquake as recently as 1959, whereas the Grand Canyon of Yellowstone is a prime example of wearing away by water on a grand scale.

Over the last 70 million years, climates have changed virtually continuously (Miller et al. 1987). The Pleistocene saw at least four major changes in climate due to alternating glacial and interglacial periods, although there may have been many important events on a shorter time scale. Over the last one million years, the earth oscillated between 90,000-year-long cold periods with ice accumulation and 10,000-year-long warm periods of ice melting (Muller and MacDonald 1997, Petit et al. 1999).

Pleistocene environments shaped the current Yellowstone landscape, flora, and fauna. Changes in the distribution of plants and animals and extinction of species were prevalent. Horses, camels, mammoths, and many other large mammals became extinct, whereas caribou, lemmings, musk ox, and other mammals retreated to the north at the end of the Wisconsin glaciation about 10,000 years before present (YBP). In addition, the abiotic changes must have caused adaptive responses—through both genetic evolution and behavioral or other plastic responses—in Yellowstone's biota. However, information to evaluate these changes is sparse or completely lacking.

PALEOBIOLOGY OF THE
GREATER YELLOWSTONE AREA

A paleobiological perspective is useful in identifying prehistoric processes that have shaped the Greater Yellowstone ecosystem (GYE), determining whether they continue to operate, and describing reasonable boundaries that may be placed on future variation.

The combination of species that simultaneously inhabit a given geographic location may be termed a community. However, communities tend not to be conserved in response to environmental change. Instead, individual species adapt and remain where they have been, or disperse at rates, times, and in directions in response to their individual tolerances to changing climatic, geological, anthropogenic, or other environmental conditions (Gleason 1926; Curtis 1955; Whittaker 1956, 1970; Davis 1976; Graham 1986). Thus, within the GYE, community composition is not stable over time scales of hundreds to thousands of years.

Many of the climate changes that occurred in the GYE region throughout the late Holocene (the past 4,000 years) were relatively small. If future climates do not have Holocene analogs (Bartlein et al. 1997), then the ranges of individual species may shift to areas that can support them outside the GYE. Although the current paleobiological record cannot document all the biotic changes associated with climate fluctuations, it does show various rates of change in climate and vegetation, some of which were extremely abrupt. Changes characteristic of the Pleistocene could happen within a human lifetime or less in the future (Alley 2000). Because the GYE is an "ecosystem island" within a larger human-dominated landscape, it is too small to accommodate environmental changes of the magnitude and frequency that were characteristic of the middle Holocene and late Pleistocene without changes in community composition, including local extinctions, greater than those seen during the past 4,000 years.

Changes in the late Quaternary (from about 14,000 YBP) vegetation and climate have been summarized by Barnosky et al. (1987) and are paraphrased here. With the initiation of deglaciation, 13,000 to 14,000 YBP, sagebrush (*Artemisia* spp.) dominated the vegetation of the subalpine forest zone. By about 11,500 YBP spruce began to colonize the area, followed by lodgepole pine, Douglas fir, and whitebark/limber pine during a period of warmer, wetter, and more stable climate (Taylor et al. 1997). By 4,500 YBP lodgepole pine became dominant with trace amounts of Douglas fir, both of which are common at lower elevations today.

Yellowstone's pre-Holocene history includes a series of diverse faunal communities, many of which have no modern-day analog. Records from the late Pleistocene are incomplete but sufficient to illustrate the degree to which animal populations can respond to changing environmental conditions; they illustrate the types of change likely to occur in the future. Because these changes are not directly relevant to conditions over the past few millennia, a more complete description is given in Appendix A.

UNGULATE USE OVER TIME AS
ESTIMATES OF POPULATION DENSITIES

Prehistoric Native Americans exploited ungulates (pronghorn, deer, moose, sheep, elk, and bison) from the time they first entered the new world in the Late Pleistocene (Frison and Stanford 1982, Kay 1990, Cannon 1992). Bison remained a stable resource for Native Americans throughout the Holocene (Frison and Bradley 1991). At some early Holocene sites, hundreds of bison have been found in association with Paleo-Indian projectile points and stone tools (Wheat 1972, Frison 1974). At sites within the GYE, however, most bone beds associated with human hunting contain fewer than five individual animals (Cannon 1992).

Unlike bison, no massive bone beds of elk have been found associated with cultural materials, and there is no evidence of artificial elk traps or communal procurement practices like those used for bison (Frison and Bradley 1991). Arguments for low population densities of elk in the past have been based on bone frequencies from archaeological and paleontological sites, journal accounts of early travelers, and historical photographs (Kay 1990, 1994, 1995; Schullery and Whittlesey 1992; Kay and Wagner 1994). Elk bones are not numerically as common at archaeological sites as pronghorn, deer, and sheep bones (Kay 1990).

The use of bone frequencies from archaeological sites as a proxy for population levels depends on the assumption that people kill and eat ungulates in proportion to their actual abundance. This assumption probably is not always valid. Anthropological studies of modern hunters show that humans, like other predators, have preferences in prey selection that are independent of abundance. Therefore, that none of the archaeological sites in the GYE yields abundant elk remains does not necessarily indicate that elk were not abundant. Problems with the relationship between bone frequencies from anthropological and paleontological sites and local ungulate abundance are further detailed in Appendix A.

PHYSICAL CHANGES FROM
PREHISTORY TO THE PRESENT

To assess the extent and nature of changes resulting from the adoption of natural regulation by the park, it is necessary to distinguish changes due to biological interactions during the past several decades from changes due to abiotic factors. We review possible abiotic factors first and then turn to biological interactions.

Climate

The climate of YNP has fluctuated for the entire period for which we have records, sometimes quite rapidly (Dansgaard et al. 1993, Alley et al. 1996). Climate histories can be developed for periods far into the past, but we focus on the period that has most directly led to the current environments in the GYE.

Glaciation: 18,000 YBP

At the peak of the last major glacial period about 14,000 YBP, large areas of North America were covered by glacial ice that had built up to a thickness of about 1.6 km over the previous 100,000 years. The continental glaciers penetrated only to the Canada-Montana border, but glaciers formed in the mountains of the GYE and spread to lower elevations and coalesced into the Yellowstone Ice Cap. Glaciers redistributed soils and sediments, widened valley bottoms, and blocked streams to form large lakes. These actions created some of the dominant landforms still present in YNP and the influence of glacial processes continues to the present. Streams draining from glaciers deposited sheets of gravel and sand in valleys below the glaciers. Some of these were later cut down to form terraces.

Holocene Climate: 10,000 YBP

Climate is much more important than geological events in shaping the biota of the northern range at the scale of thousands of years. Seasonal extremes in temperature or moisture, rather than mean annual values, are probably the more important limiting factors for organismal distribution. Areas with the same average temperature and precipitation but highly different variation and seasonal patterns support very different biotas. Several important climate shifts strongly influenced Yellowstone during the Holocene. The warmest climates occurred in the early Holocene (7,000 to 9,000 YBP). The Medieval Warm Period dates to 1100 to 1300 AD (900 to 700 YBP) and it was manifest in droughts in the northern plains of the United States (Laird et al. 1996, Woodhouse and Overpeck 1998).

Of particular interest for Yellowstone are the effects of the Little Ice Age, which lasted from 1450 to 1890 AD in two main pulses and appears to have been global (Crowley and North 1991). Temperatures averaged 1 to 1.5° C cooler (Crowley and North 1991) and the climate tended to be moister than it is today. During the Little Ice Age, glaciers advanced worldwide. Within this period, there were periods of warm and dry and cool and wet (second pulse of the Little Ice Age, 1860 to 1910) conditions, which affected the growth and distribution of plant species (Whitlock et al. 1995). In the GYE, this record is best documented by tree rings.

The stability of the present climate appears to be abnormal compared with recent millennia, although the climates of the past 200 years have included the warmest and coldest periods of the past 4,000 years (Bradley 2000).

On a decadal time scale, Yellowstone's climate is measurably influenced by El Niño-Southern Oscillation (ENSO) patterns, primarily through the creation of extremes in precipitation and temperature. These extremes themselves probably play a major role in shaping the GYE's environment and biota. However, although climate variability is subject to ENSO periodicity, overall annual precipitation shows no statistically significant trend over the past 100 years (Balling et al. 1992a).

Fire in Yellowstone and the Northern Range

Recurrent wildfire profoundly influences fauna, flora, and ecological processes in the northern Rocky Mountains (Habeck and Mutch 1973; Houston 1973; Loope and Gruell 1973; Taylor 1973; Wright and Heinselman 1973; Wright 1974; Arno 1980; Romme and Knight 1981, 1982; Romme 1982; Knight 1987, 1996; Romme and Despain 1989; Despain 1990; Turner et al. 1997). Although small fires occur frequently, total area burned is dominated by a few, very extensive fires (Johnson and Fryer 1987, Romme and Despain 1989, Johnson 1992). Meyer et al. (1995) found weak evidence of fire as long ago as 7,500 and 5,500 YBP and strong evidence for substantial episodes 4,600, 4,000, 2,500, 2,100, 1,800, 1,200, and 850 YBP. Between 9 and 12 maxima can be identified in the record between 2,000 YBP and the present, including the 1988 fire. Major fire events appear to have occurred recently at 100- to 300-year intervals (Romme 1982, Romme and Despain 1989) and at various frequencies throughout the Holocene (Millspaugh and Whitlock 1995). Within the ungulates' summer range, large, infrequent fires create vegetation mosaics that dominate the landscape until the next extensive fire.

On the northern range, tree-ring evidence and fire scar data indicate that 8 to 10 extensive fires occurred in the area during the last 300 to 400 years, which suggests that fires burned the winter range at intervals of 20 to 30 years before European settlement (Houston 1973, Barrett 1994).

Climate plays an important role in fire frequency and extent. Most large North American twentieth century fires have been associated with persistent high-pressure ridges or dry La Nia phases of ENSO (Bessie and Johnson 1995, Swetnam and Betancourt 1998). Within Yellowstone, the area burned correlates with a trend toward increasing late-winter aridity since 1895 (Balling et al. 1992a, 1992b). However, the potential for climate-induced changes in fire frequency and extent in the Northern Rocky Mountains (Flannigan and Van Wagner 1991, Romme and Turner 1991, Bartlein et al. 1997) underscores the importance of understanding the effects of extreme events and of considering long-term disturbance dynamics when considering alternative management strategies.

Lower elevation communities, such as those found on the northern range where fires were formerly frequent, have been altered by long-term fire suppression. Fire suppression was discontinued in 1972 in Yellowstone, after which lightning-caused fires were again allowed to burn. As with many other crown fire-dominated ecosystems, YNP is usually considered a nonequilibrium landscape (Romme 1982, Sprugel 1991, Turner and Romme 1994).

No large fires occurred during the twentieth century until those of 1988, the largest since the park's establishment, affected more than 321,000 ha in YNP and the surrounding area and burned approximately 36% of the park (Schullery 1997). These fires were primarily the result of unusually prolonged drought and high winds (Renkin and Despain 1992, Bessie and Johnson 1995) and were consistent with the earlier pattern of punctuated episodes of extreme fires followed by long periods of small fires (Millspaugh and Whitlock 1995). Reconstructions suggest that the last time a fire of this magnitude occurred in Yellowstone was in the early 1700s (Romme and Despain 1989), which makes the 1988 fires unusual in size (Christensen et al. 1989, Turner et al. 1994a,b,c). However, fire suppression probably had only minimal influence on the extent and pattern of the 1988 fires (Romme and Despain 1989, Barrett 1994).

The effects of fire on wintering ungulates change through time following the fire. Initially, fire consumes aboveground plant biomass and reduces winter forage supply, which is not regenerated until the following summer. This reduction in forage can increase ungulate mortality during the first winter after a major fire (Singer et al. 1989, Wu et al. 1996), an effect that is magnified during more severe winters. In subsequent years, primary productivity may be stimulated, resulting in improved forage quantity and palatability (Harniss and Murray 1973, Gruell 1980, Hobbs and Spowart 1984, West and Hassan 1985, Coppock and Detling 1986). Some plant communities on the northern range showed substantial increases in forage abundance and nutrient content in response to burning (Wallace et al. 1995, Tracy and McNaughton 1997). Ungulates on the northern range preferentially used herbaceous plants in burned areas of the landscape (Pearson et al. 1995). Boyce and Merrill (1991) hypothesized that this fire-enhanced forage base in burned forests (i.e., increased herbaceous ground cover) might enhance ungulate recruitment and population size for several years following the 1988 fires. However, enhancement of forage production in grassland areas was no longer detectable approximately 5 years after those fires (Singer and Harter 1996).

ENSO is not directly correlated with YNP fires because the quantity and ignitability of herbaceous fuels are strongly influenced by local weather and time since the most recent catastrophic burns. Nevertheless, if the GYE's climate is warming, and ENSO events are becoming more frequent, then fire can be expected to be a more dominant process in the future (Millspaugh and Whitlock 1995).

Fire also has geomorphological consequences, because vegetation is removed by burning, and storm runoff occurs as overland flow. Charcoal layers that commonly accompany or follow fires appear in expanded alluvial sediments (Meyer et al. 1992). Erosion continues in subsequent years, even after revegetation, because deep incision contributes to slumping of fire basins into channels. The clearest demonstration of the impact of fire on erosion is the aftermath of the 1988 conflagration (Meyer et al. 1992, 1995). Meyer et al. documented major erosional debris flows that extended over several hundred meters in the Slough Creek and Soda Butte Creek drainages of northeastern YNP. Meyer et al. (1992) proposed an idealized scenario to explain how wet periods during the Holocene created widened and sinuous stream beds, and dry periods with burns led to erosion of steeper slopes and aggradation of alluvial fans and incised stream courses. Whether the processes are captured correctly by this scenario, clearly events of this magnitude reshape the character of streams, both large and small, and constitute a greater influence than elk browsing on willows and other riparian woody vegetation.

Such major forces affect not only stream morphology and seasonal flow but also the likelihood of persistence of beavers with their modification of stream flow, water-table height, and extent of riparian vegetation.

This evidence suggests that relatively minor climatic changes in the late Holocene could have caused major shifts in fire regimes, alluvial processes, and resulting morphology and vegetation of valley floors.

Stream Flow and Channel Morphology

Although many factors influence runoff, typically there is a strong positive correlation between precipitation and stream flow. No long-term trend in stream flow at the Corwin Springs gage on the Yellowstone River just north of the park is apparent through the 91-year record from 1908 to 1999 ($p = 0.259$). Although other studies have shown a decline in precipitation over the same period, this is true only for late winter measurements, which accounts for a minority of annual precipitation (Balling et al. 1992a). The proportion of snowmelt that contributes to stream flow is high relative to that of rainfall, much of which is absorbed in the soil and transpired. Like stream flow, snowpack showed no significant downward trend ($p = 0.40$) but was highly subject to ENSO events ($p = 0.001$). During an ENSO, precipitation at YNP is much greater than normal, which causes major changes in stream courses.

It is usually presumed that fire increases surface runoff and stream flow by destroying vegetation, thereby increasing the likelihood of erosion and bank instability. After the 1988 fires, however, no significant increase in stream flow was observed in the Yellowstone River.

Stream-course changes during the flood years of 1996 and 1997 were the greatest observed since 1954, primarily the result of flood flows (Mowry 1998). From 1954 until 1987, a period of relatively constant stream flow, stream channels narrowed, especially in areas where elk wintered, a process described by Lyons et al. (2000). From 1988 to 1997 there was increased channel instability, and most of the changes occurred in the exceptional flow years of 1996 and 1997. Most stream-bank erosion during these extreme runoff years occurred higher in the drainage—where there were few wintering elk and the streamside willows were robust—suggesting that increased erosion was primarily due to hydrological variables acting on high-roughness riparian vegetation that stabilized the stream bank during high-flow, nonflood years and not to ungulate activity. This conclusion does not explain the long-term development of low-roughness vegetation (i.e., vegetation with little resistance to stream flow power) in areas with heavy winter elk use. However, these differences do explain Mowry's (1998) finding that there was a greater change in stream reaches with willow-covered banks than in those with grass-dominated banks. Shifts in stream morphology associated with increased stream flow are expected because the stream slope is near the transition of meandering and braided stream courses based on the established relationship between channel slope and bankfull discharge (Leopold and Wolman 1957).

Climate, vegetation, and stream-course relationships interact in multiple ways, but it is difficult to assess the role of extreme events because they happen so infrequently. Nevertheless, it is apparent that both physical forces and browsing and streambank trampling by ungulates contribute to changes in streams, but the data are not sufficient to sort out their relative importance in causality.

Upslope Soil Erosion

Stream-bank erosion is only one of the issues raised about management of the northern range. Critics of the natural regulation policy in the northern range claim that degradation of upland slopes by overgrazing results in accelerated soil erosion (YNP 1997). One indication of the level of erosion off the slopes is the accumulation of sediments in depressions or lakes. Engstrom et al. (1991) studied several lakes on the northern range, looking at pollen changes and abnormal sediment deposition patterns. They concluded that their "investigation of the sedimentary record does not support the hypothesis that ungulate grazing has had a strong direct or indirect effect on the vegetation and soil stability in the lake catchments or on the water quality of the lakes." A review of their data, however, shows many, but not all, lakes with increasing sediment accumulation starting after the beginning of the twentieth century. Most of the study lakes near the confluence area of the Yellowstone River, Lamar River, and Slough Creek showed sediment increases. One study lake in this area did not. Interpretation of these data might vary by investigator, and the magnitude of the increases may be "normal," although sediment accumulation in the nineteenth century seems to be comparatively stable when compared to the twentieth century.

Changing Land Use in the Greater Yellowstone Area

The GYE includes not only YNP but also most of the adjacent lands at an elevation above 1,524 m (5,000 ft). These surrounding lands are used for ranching, agriculture, and recreation. The population of Park County, north of and adjacent to the northern range of YNP, is growing faster than the populations in most Montana counties.

In 1880, the population of Park County was 200, but after the arrival of the Northern Pacific Railroad in 1881, it grew to 6,900 in 1890. Population growth has continued and today the county (population about 16,000) is dotted with many new developments and small ranches with increased fencing, as well as many semi-urban areas (Park County 2001).

This area was reported to be the ancestral wintering range of the northern range elk herd before YNP was established (Graves and Nelson 1919). The influence of development and farming was noted by Wylie (1882,[1] 1926[2]) as early as the late 1800s and early 1900s. He said, "The buffalo, deer, and elk were accustomed to living on this plateau during the summer. In winter they migrated to lower and warmer regions outside the park area until settlements of farmers in the country surrounding the park made it impossible for them to use their long used winter homes." All these factors fragment habitat and impede ungulate movements and access to foraging areas. This alteration of the landscape outside of YNP may have as great a potential to affect ungulate populations, their behavior, and the use of vegetation as do changing climatic conditions and reintroduction of wolves. Little is known, however, about the relative importance of each of these factors or their interrelationships.

---

[1]Wylie, W.W. 1882. Yellowstone National Park the Great American Wonderland: A Complete Hand or Guide Book for Tourists (unpublished material).

[2]Wylie, W.W. 1926. History of Yellowstone Park and the Wylie Way Camping Company (unpublished material).

3

Present Conditions: Vegetation

Vegetation of Yellowstone's northern range, a mosaic of different forest and nonforest communities, is the result of interactions among many environmental factors. To understand how management decisions may affect vegetation of the northern range, the committee reviewed the present conditions of the major vegetation types. In this chapter, the current state of a type of vegetation—for example, sagebrush or aspen communities—is described, followed by a discussion of how the driving variables may affect these conditions. In several cases, where changes in driving variables may alter ecosystem components, the nature of these changes and their consequences is discussed. Recent modification of environmental drivers is discussed to help explain the significance of recent changes of the ecosystem components. Although this chapter emphasizes the dominant plants, such as sagebrush or aspen, the concern over loss or degradation of these systems is not only for the dominant species, but also for the biodiversity—plants, animals, fungi, and microbes—that they support.

UPLAND SHRUBLANDS AND GRASSLANDS
OF THE NORTHERN RANGE

Shrublands

Big sagebrush-Idaho fescue is the most abundant sagebrush-grassland type. It occurs on sites with thin cobble soils to well-developed loams, generally at elevations of 1,800 to 2,400 m within the 40- to 75-cm precipitation zone. It is distributed throughout the park but is most common in the Gardner and Lamar River drainages (Despain 1990). The habitat type is dominated by mountain big sagebrush (*Artemisia tridentata* ssp. *vaseyana*), although Wyoming big sage (*A. tridentata* ssp. *wyomingensis*) may also be present. Identification of big sagebrush subspecies is particularly important because of differences in palatability and preference to ungulates. Idaho fescue (*Festuca idahoensis*) dominates the understory with *Agropyron spicatum* and *Koeleria macrantha* also present. Forbs (broad-leaved herbaceous plants), such as *Geum triflorum*, are abundant.

Primary production (the amount of carbon fixed by photosynthesis) varies widely in big sagebrush-Idaho fescue habitat depending on rainfall and temperature (Mueggler and Stewart 1980). A 50% difference in production may occur on any given site over a 3-year period. Production varied across the type from 560 kg/ha (Mueggler and Stewart 1980) to 1,610 kg/ha with grasses contributing 21% to 42% of the production, forbs 38% to 56%, and shrubs 10% to 41%. Between 88% and 98% of the shrub production is from big sagebrush.

Big sagebrush-Idaho fescue habitat, which is heavily grazed in winter by ungulates, and the grassland habitat type (Idaho fescue-bearded wheatgrass) account for slightly more than half of all the nonforested vegetation in the park and on the northern range (Houston 1982). These two types probably furnish most of the forage for the large number of grazing animals in the park (Despain 1990).

Wyoming big sagebrush-bluebunch wheatgrass (*A. spicatum*, now *Pseudoroegneria spicatum*) habitat type occurs in the Gardner River canyon in small areas on southern and western slopes, often between big sagebrush-Idaho fescue and other grasslands on ridgetops and upper slopes. It occurs on shallow to moderately deep soils formed over several parent materials.

Mountain big sagebrush is the dominant shrub, although basin big sagebrush (*Artemisia tridentata* ssp. *tridentata*) may occur on deeper soils in drainages. Low shrubs, such as *A. frigida* and *Gutierrezia sarothrae*, are usually present. In addition to bluebunch wheatgrass, other conspicuous grasses include *K. macrantha, Poa secunda,* and *Stipa comata*. Production varies between 670 and 1,120 kg/ha with high variability between sites but not between years (Mueggler and Stewart 1980). This type is heavily grazed in winter by ungulates in the Gardiner area. Big sagebrush receives enough browsing to reduce the size of its canopies.

In the Gardiner area, Wambolt and Sherwood (1999) also describe a Wyoming big sagebrush-bluebunch wheatgrass (*A. spicatum*) habitat type, as did Mueggler and Stewart (1980) and Houston (1982). Associated species include sprouting shrubs like rubber rabbitbrush (*Chrysothamnus nauseosus*), green rabbitbrush (*Chrysothamnus viscidiflorus*), and gray horsebrush (*Tetradymia canescens*). Prairie junegrass (*K. macrantha*) and Sandberg bluegrass (*P. secunda*) are also common.

Grasslands

Idaho fescue-bearded wheatgrass (*Agropyron caninum*) habitat type is a highly productive mesic grassland with high species diversity. It occurs on gentle slopes at elevations of 2,000 to 2,600 m, within the 46- to 76-cm rainfall zone. It is dominated by grasses but contains a higher proportion of forbs (30% to 70%) than other western Montana habitat types. It has a short growing season and is used by native ungulates in winter (Houston 1982).

Idaho fescue-Richardson's needlegrass (*Stipa richardsonii*) habitat type generally occurs at elevations of 1,100 to 2,100 m on gentle slopes and deep soils. It is a moderately mesic and productive grassland type dominated by *Festuca idahoensis, S. richardsonii, Danthonia intermedia, Stipa occidentalis*, and *G. viscosissimum*. This grassland is summer range for sheep or cattle, and it receives substantial winter grazing by native ungulates.

Idaho fescue-bluebunch wheatgrass is the most common xeric (Houston 1982) or moderately mesic (35 to 50 cm of precipitation) grassland type in the Greater Yellowstone Ecosystem (GYE) (Mueggler and Stewart 1980). It is found on intermediate mountain slopes at elevations of 1,400 to 2,300 m and occurs on a wide variety of parent materials.

Other grasses include *K. macrantha, P. sanbergii*, and either *S. comata* or *S. occidentalis*. Forbs cover from 10% to 60% of the area and include *Achillea millefolium, Antennaria rosea, Arenaria congesta*, and possibly *Phlox hoodii*. Medium shrubs such as *A. tridentata* and *C. viscidiflorus* are occasionally present. Annual primary production is highly variable depending on the weather. The grasses are used by elk and deer at the lower elevations for winter range and by pronghorn year-round. At the highest elevation, this type is summer range for elk and deer. At middle elevations, it is used as spring and fall range by all ungulates and as winter range by bighorn sheep and mountain goats.

Bluebunch wheatgrass-Sandberg bluegrass, or *A. spicatum-P. sanbergii,* is usually found between 900 and 1,800 m, especially on gravelly soils on steep southern slopes. It is a moderately arid type in the 35- to 50-cm precipitation zone. Shrub and forb cover is low, and rhizomatous grasses are generally absent.

Needle-and-thread-blue grama (*S. comata-Bouteloua gracilis*) habitat type is usually found on broad alluvial benches and valley floors. Houston found this grass type in Yellowstone National Park (YNP) in the boundary line area, upstream on the Yellowstone River to the Black Canyon. It generally occurs below 1,500 m and is the driest grassland habitat type (20 to 35 cm precipitation). The type is floristically simple, containing grasses and a low cover of forbs and shrubs. Needle-and-thread grass is a bunchgrass that dominates late seral stages of the community but decreases under heavy grazing pressure. Blue grama, the other dominant sod-forming grass in the community, increases under heavy grazing.

The terms decreaser, increaser, and invader refer to a plant's response to grazing (Dyksterhuis 1949). Decreaser plants are most preferred by grazing animals and with continued heavy grazing are the first kinds of plants to decline in cover in a community. Increaser plants initially increase in cover in a community under ungulate grazing pressure because the preferred decreaser plants are declining and opening up space for increasers to grow. Eventually, as heavy grazing pressure continues, the increaser plants also decline, opening up sites for invader species of low palatability and generally low nutritional value. Common shrubs that increase with overgrazing include *A. frigida, G. sarothrae,* and *Opuntia polycantha.*

Factors Influencing Present Conditions of Sagebrush and Grasslands

Sagebrush: Ungulate Use

Big sagebrush is a particularly important food plant for several Yellowstone ungulates, especially in winter. Consequently, lower sites where there is little snow or where snow does not deeply cover shrubs are heavily grazed. During other seasons use is less, although big sagebrush may be an important source of protein for elk during the gestation period and in summer because grasses alone do not meet their protein needs (Wambolt et al. 1997).

Not all ungulates in the northern range use big sagebrush to the same extent and not all subspecies of big sagebrush are equally used by the ungulates that feed on it. Big sagebrush is an important component of elk, mule deer, and pronghorn diets but not of bighorn sheep, bison, and mountain goats (Houston 1982). Mule deer and elk strongly prefer mountain big sagebrush to Wyoming big sagebrush and basin big sagebrush (Wambolt 1996). Wyoming and basin big sagebrush are much preferred over black sagebrush (*A. nova*), but during severe winters, all subspecies of sagebrush are browsed.

Ungulate browsing in low-elevation sagebrush sites near the park boundary has resulted in significant negative effects on big sagebrush (Wambolt 1996). In some cases, up to 91% of the leaders were removed and unbrowsed plants had higher productivity (45 g per plant) and seed-head production (60.3 seed heads per plant) than browsed plants (10 g per plant and 0.08 seed heads per plant) (Hoffman and Wambolt 1996). Up to 35% of plants were killed between 1982 and 1992, and many plants that survived had high percentages of dead crown (Wambolt 1996). Wambolt's exclosure work (1998) demonstrates elk-induced decreases of sagebrush even in areas where there were no other ungulates. Wambolt and Sherwood (1999) came to similar conclusions. However, according to the National Park Service (YNP 1997), on 97% of the northern Yellowstone winter range sagebrush is stable or increasing and only 3% of the land shows sagebrush decline. In general, less browsing damage is observed at higher elevations (Singer and Renkin 1995, YNP 1997). Thus, elk appear to affect sagebrush at lower elevations—including the 3% of the winter range described by NPS as having declining sagebrush—but not at higher elevations.

Grasslands: Ungulate Use

Grasses are important components of the diet of most YNP ungulates except pronghorns and moose (Singer and Norland 1994). The importance of grasses in the diets of most ungulates on the northern range of YNP was shown by Singer and Norland (1994) through microhistological analyses on feces, comparison with earlier published work based on rumen analyses, and, for bighorn sheep, examination of feeding sites. Mean percentage diet compositions assessed by these methods were, as follows: elk, 75% to 79%; bison, 53% to 54%; mule deer, 19% to 32%, pronghorn, 10% to 4%; and bighorn sheep, 65% to 58%. Bison also included a high proportion of sedges in their diets (32% to 56%).

As in many other grasslands, ungulates in YNP move nonrandomly over the landscape, feeding preferentially on grasses that are at particular stages of development (Frank et al. 1998). Yellowstone elk migrate elevationally as grasses produce new growth in spring (Houston 1982, Frank and McNaughton 1992). Some areas are intensively grazed but they recover as animals move to other patches. Timing of feeding is critical because feeding on vegetative material can have less impact than removal of growing points or reproductive structures. Despain (1996) compared one exclosure with the surrounding area. He found that elk fed heavily on the highly palatable bluebunch wheatgrass but moved off before the grass flowered so that there was little difference between exclosures and the surrounding areas in the total amount of green biomass of all species at the end of the growing season.

The grazing and migration pattern in the northern winter range results in modest spring and summer grazing on the lower ranges that receive heavy winter pressure and more intense grazing at higher elevations as snow recedes and green-up occurs (Singer and Harter 1996).

There are visually apparent effects of grazing on YNP grasslands. The question is whether those effects are signs of damage induced by feeding populations that exceed the carrying capacity of those rangelands. NPS perspective at YNP (YNP 1997) is that there is a perceptional problem—observers who see YNP rangelands make comparisons with commercial livestock rangeland and interpret the differences to indicate overgrazing (Coughenour and Singer 1991, 1996a).

Grasses are generally adapted to grazing and may even respond positively to appropriate grazing levels (Huff and Varley 1999). Grazing in YNP may cause enhanced plant protein (Singer 1996) and nitrogen content (Coughenour 1991, Mack and Singer 1993), and grazed plants may produce taller leaf and seed stalks (Singer and Harter 1996). Also, it is possible that animal movements, deposition of urine and feces, and the physical effects of hooves combined with plant responses to grazing could result in dense, short grass stands of enhanced above-ground growth (Frank and McNaughton 1992). Plant diversity on grazed sites is often higher than on ungrazed or heavily grazed sites (Wallace et al. 1995). Some authors have suggested that there are alarming decreases in plant diversity on Yellowstone's northern range, implying that overgrazing is occurring on some sites (Wagner et al. 1995).

Grazing in the northern range does not appear to reduce root biomass (Coughenour 1991) or soil moisture content, even though there is an increase in soil bulk density (Lane and Montagne 1996). Changes have been reported in forb biomass (Singer 1995) and soil nutrients (Lane and Montagne 1996).

Confounding Factors: Fire, Pocket Gophers, and Nonnative Plants

Three other factors affect shrublands and grasslands in the northern range, but they have received little study.

Usually, fire is not common in grass or big sagebrush communities in the northern range (Despain 1990), but unusual increases in sagebrush since the 1870s (Houston 1982) may have caused sufficient fuel buildup to carry fires in this fire-sensitive vegetation type. After fire, numerous seedlings may establish, although there is little evidence that these plants survive and reproduce, especially for mountain big sagebrush (Wambolt et al. 1999). Possibly, the lush, young greenery draws ungulates to the site, increasing herbivory and further decreasing sagebrush population recruitment (Wambolt et al. 1999). Fire has fewer negative effects on grasslands and may stimulate community renewal.

Pocket gophers (*Thomomys talpoides*) are common, active, fossorial rodents that move masses of soil wherever there is sufficient below-ground consumable biomass to support them. Their digging activities, and those of bears, create patches in YNP grasslands with high densities of forbs that replace grasses and add to community diversity (Despain 1990). Pocket gophers can alter the structure of a community significantly and change the time-course of succession (Chase et al. 1982). Activities of elk and other ungulates might affect *Thomomys* populations by altering vegetation cover and soil compaction and consequently indirectly influence vegetation characteristics. If this had occurred, it would complicate the assessment of ungulates' effects on vegetation, but we are not aware of relevant data for the northern range.

Nonnative plants are abundant in Yellowstone's northern range, especially in big sagebrush and grassland habitats. Three grass species, timothy (*Phleum pratense*), crested wheatgrass (*Agropyron cristatum*), and cheat grass (*Bromus tectorum*), occur in the area and might alter, to an unknown degree, ecosystem productivity, susceptibility to fires, and ecosystem nutrient dynamics as well as other integrated measures of ecosystem processes (Huff and Varley 1999).

Conclusions

Not enough data are available for the committee to evaluate the claim that abiotic factors in the northern range are currently more influential than biotic factors on vegetation (YNP 1997, Frank et al. 1998), or the reverse. Certainly, many processes, especially those in the soil, are strongly mediated by animals in YNP (Frank and Groffman 1998). The available data indicate that over short periods (decades), browsing and, in some areas, grazing have caused declines in plant populations and productivity.

Sagebrush

Big sagebrush at higher elevations in the northern range appears to be at relatively high abundance and vigor. These areas are important for ungulates during the nonwinter portions of the year. Because snow is usually deep and plants are relatively protected from elk browsing, these areas do not show sagebrush decline and may even show increases. They are not obviously of immediate management concern.

Lower-elevation big sagebrush stands, which have been very heavily used by elk, are decreasing in density and productivity, especially near the northern park boundary. Those sites are a critical winter range for a variety of other ungulates, especially pronghorn. It appears that, without extensive and intensive management to offset the damage done by elk browsing and grazing, the sites will continue to be degraded as resources for pronghorns and other ungulates (Wambolt and Sherwood 1999), especially near the northern boundary of the park.

Grasslands

Grasslands do not appear to have been altered as much by grazing as low-elevation shrublands have been by browsing. However, the few comprehensive reviews of the literature do not factor in the amount of biomass or other integrated measures of ecosystem characteristics contributed by nonnative species. The few studies available do not indicate that biodiversity is declining or that these systems are near a threshold value for some characteristic that is critical to any ecosystem process that currently appears to be within normal, long-term variations of the system. However, the committee would have more confidence if there were more data and analyses available.

FOREST TYPES ASSOCIATED WITH THE NORTHERN RANGE

YNP forests range from lower elevation woodlands through dense forest to timberline woodlands of whitebark pine and spruce and fir krummholz.

Knight (1994) describes seven forest and woodland types in Wyoming, six of which are found in YNP. The seven include limber-pine woodland, ponderosa-pine forest, Douglas-fir forest, aspen forest, lodgepole-pine forest, spruce-fir forest, and whitebark-pine woodland. There are many associations within these forest types, based on the composition of dominant and subordinate understory plants. Despain (1990) uses these associations to describe many forest habitat types. Of the seven types described by Knight, only ponderosa-pine forests are not found in YNP, and limber-pine woodlands are not common. The lack of ponderosa pine is considered to be caused by the predominance of rhyolitic soils on the Yellowstone Plateau, soils that create water stress conditions too extreme for ponderosa pine in the elevational zone where ponderosa pine might have established. The area may also lack the higher summer precipitation and warmer and longer growing season temperatures required by ponderosa pine. Of the six YNP forest or woodland types described by Knight (1994), aspen forests are discussed separately in this chapter because of the importance of their growth and reproductive response to changing environmental conditions in the northern range.

The forests of northern YNP and adjacent areas exist in a mosaic of forests and meadows (or parks). The causes for this mosaic mostly relate to moisture availability, whether influenced by soils, topography, or other factors (Patten 1963). The forests of YNP are continuously changing as stands mature and external factors, such as long-term climatic changes (Whitlock 1993), cause decline or loss of existing stands. Additionally, because of natural or anthropogenic environmental changes, such as climate change or fire suppression, many nonforest areas have been invaded by trees (Patten 1969, Jakubas and Romme 1993). Historical photograph comparisons show that many slopes throughout the northern range have more conifer forests now than in the past (Meagher and Houston 1998). These photographic comparisons also show a decline of aspen stands throughout the area, a phenomenon also seen in remotely sensed data that show aspen changes inside and outside northern YNP (Ripple and Larsen 2000a). Invasion by conifers of sagebrush and other nonforested areas continues to occur throughout the GYE. Sagebrush areas are becoming forest (Patten 1969), and forests are invading subalpine meadows (Jakubas and Romme 1993).

The most common forest type in the lower elevations of the northern range is dominated by Douglas fir (*Pseudotsuga menziesii*). It occurs below the lodgepole pine zone (1,800 to 2,300 m) in dense stands on cooler sites and sparse stands often mixed with Rocky Mountain juniper on drier or warmer sites. Douglas fir develops a thick bark, which makes mature trees relatively fire tolerant. In the Yellowstone and Lamar River valleys (a major portion of the northern range), Douglas fir is the most common tree, with snowberry a common understory shrub in warmer sites and pine grass common in cooler sites. Aspen and lodgepole pine are often associated with Douglas fir in these areas. Douglas-fir forest with shiny-leaf spirea and other short woody shrubs as understory is common in the northern range on upper slopes and ridges (Despain 1990).

The extensive zone above Douglas-fir forest in the northern range is dominated by lodgepole pine. Although Despain (1990) described a few pure lodgepole-pine habitat types in YNP, none were in the northern range. However, using a cover classification, he described many cover types based on different successional stages of lodgepole forests with different understory recovery phases of climax species (e.g., subalpine fir). Knight (1994), however, described the lodgepole-pine forest as the most common forest type in Wyoming, occurring in northern Wyoming from 1,800 to 3,200 m. He pointed out that, although lodgepole pine is primarily a fire successional species, climax lodgepole pine can occur on cool, nutrient-poor sites where other Rocky Mountain conifers cannot survive.

The forest zone above the Douglas-fir zone includes other subalpine species such as subalpine fir, Engelmann spruce, and, at higher elevations, whitebark-pine woodlands. Spruce, fir, and sometimes whitebark pine form stunted krummholz "forest" stands on the ecotone between forest and alpine communities. In most cases, the krummholz is on exposed ridges or rocky outcrops. The elevation of these woody communities is well above the northern range especially the northern winter range.

Factors Influencing Present Conditions of
Northern Range Forests

The present condition of forests on the northern range has been determined primarily by changes in management of fire and ungulates

Fire

Fire management policy has changed over the past several decades in YNP as well as in surrounding national forest wilderness areas. Before the 1970s, all fires were extinguished regardless of their location or intensity.

During the 1970s, the benefits of fire were recognized, and fires that were unlikely to damage human activities or structures were left to burn. This was very successful as most fires in the park during the decades leading up to 1988 burned a few to several hundreds of hectares in Yellowstone's conifer forests, and the forest mosaic normally formed by disturbance processes was gradually returning to a more "natural" landscape. Fires did not occur in the sage-brush-grasslands during this time. (The fires of 1988 would have occurred even if decades of controlled burns had preceded them.)

Forests of the northern range are a mosaic of burned and unburned stands, most of which burned in 1988 (Despain et al. 1989). Before 1988 the forests were pure lodgepole-pine or mixed lodgepole-Douglas-fir and lodgepole-spruce-fir forests (Keigley 1997a). Most of the burned forests are recovering as nearly pure stands of young lodgepole pine.

Ungulate Use

Many forest stands in the northern winter range and in the upper Gallatin River drainage are heavily browsed and highlined (i.e., a browsing pattern on trees caused by ungulates removing foliage and live twigs as high as they can reach, thus creating a high line usually a few meters above the ground) (Kay 1990). Stunted conifers may not be browsed during mild winters and thus grow into branched trees (see following paragraph for analysis of tree archi-tecture). All conifer species are browsed. Spruce, fir, and Douglas-fir trees are highlined, and adventitious branches, which grow on tree stems, are also browsed. In the upper Gallatin River drainage near the YNP boundary, trees are highlined throughout most of the area, but highlining decreases or disap-pears several miles south of the park boundary at higher elevations or several miles north of the boundary, where ungulates migrate but few overwinter (committee observations). The forest-ungulate interaction found on the Gallatin may represent a microcosm of the northern winter range. This interaction demonstrates the influence of ungulates on the forest under winter conditions where ungulates now stay at higher elevations in areas that once were primar-ily summer range (Patten 1963).

Branch architecture and growth form have been used for determining the intensity of browsing of shrubs and trees (Keigley 1997a). Young trees branch after the terminal shoots are browsed and then may grow into branched rather than single-stem tall trees if browsing pressure is reduced. Dating of the origin of branching and other tree-architectural anomalies have been tied to periods of elevated ungulate numbers, while periods of "release" of terminal branches to form tall trees appear to correspond with periods when ungulate numbers were low (Keigley 1998).

Weather and Hydrology

Changes in forest patterns may be caused by altered hydrological conditions. Certainly, some years in the decade of the 1980s had warmer, drier weather than the decades preceding it, and apparently than the decade succeeding it. Consequently, when temperatures were warmer and winds were greater than normal, small fires became big fires and the extensive fires of 1988 occurred. Such cyclical changes in climate cause short, sporadic changes in many variables that affect forest ecosystems and therefore may temporarily establish external conditions for change. These cycles are typical; thus, disturbances and changes in forests of the northern range as influenced by hydrological fluctuations are normal. How hydrological cycles influence other factors such as ungulate behavior may be an important compounding effect in determining causes for alterations of the forests and individual trees within the forests and woodlands of the northern range.

Diseases and Infestations

Disease and infestations of insects and parasites are major factors that modify the forests of the YNP and the surrounding area. Several infestations of native species such as spruce budworm and pine-bark beetle have killed many hectares of forest, sometimes producing barren slopes, whereas at other locations only individual trees die. Reduction of tree health by these infestations, especially pine-bark beetle on lodgepole pine, makes trees susceptible to other pathogenic epiphytes such as dwarf mistletoe.

Recently, invasion of nonnative white-pine blister rust into the Rocky Mountains has caused mortality in whitebark pine and limber pine (Kendall and Schirokauer 1997, Kendall and Asebrook 1998). Proximity of *Ribes* spp., the secondary host for the fungus, as well as appropriate environmental conditions are necessary for blister rust to invade pine stands. In areas where blister rust infection is very high, such as Glacier National Park, conditions are moist; in YNP, drier conditions tend to retard the spread of the rust (Kendall 1998,

Kendall and Keane 2001). Elevation also plays a role in amount of infection, as whitebark-pine stands at higher elevations, and perhaps greater distances from *Ribes* communities, show less infection and mortality than lower-elevation stands. White-pine blister rust infection continues to expand in the GYE; if whitebark pine is lost in that area, their seeds, a major food source for grizzly bears, will also be lost (Kendall 1983, Mattson et al. 2001).

Conclusions

Forests associated with the northern range form a mosaic with nonforested areas. Over the past century, probably because of fire suppression, many non-forested areas have been invaded by trees, converting many hectares of the northern range into savanna-type forests or closed-canopy forest stands. These forests have been browsed by ungulates to a limited extent, mostly where ungulates are short of winter food. This use does not appear to significantly affect forest advance or the present distribution of the forests in the northern range.

ASPEN COMMUNITIES OF THE NORTHERN RANGE

Aspen (*Populus tremuloides*) is the most widely distributed native North American tree species (Fowells 1965) and an important component of landscapes in the intermountain west. Aspen is the only native upland deciduous tree that occurs in YNP. Aspen stands support high numbers and diversity of breeding birds (DeByle and Winokur 1985) and provide habitat for other wildlife. They are an important source of forage for browsing ungulates, especially during winter (Olmstead 1979). Aspen stands also are considered prime areas for livestock grazing and can be extensive enough to provide a quality watershed and attractive scenery (DeByle and Winokur 1985). However, many aspen stands throughout the west, including YNP, appear to have declined during the twentieth century as old trees died and little recruitment took place. The causes and consequences of this decline have received considerable discussion (Krebill 1972, Loope and Gruell 1973, Schier 1975, Hinds and Wengert 1977, Olmstead 1979, Bartos and Mueggler 1981, Hinds 1985, Boyce 1989, Kay 1990, Bartos et al. 1994, Romme et al. 1995, Baker et al. 1997). Like other species of the genus *Populus* (poplars), aspen are single-trunked, deciduous trees that spread clonally by means of root-borne sucker shoots (Eckenwalder 1996). They are among the fastest-growing temperate trees, and their shoots continue to grow after bud-burst by initiating, expanding, and maturing leaves throughout the growing season. Aspen is dioecious, having separate male and female trees that flower before leaf emergence in the spring. The seeds are born on catkins and are wind dispersed, often over long distances. Seeds remain viable for only a few weeks after their release in late spring (Moss 1938). Aspen usually regenerates via vegetative suckering from the residual root system after a disturbance that kills the mature trees. These adventitious shoots grow rapidly and are supported by the parental root system for growth for at least the first 25 years (Zahner and DeByle 1965). The new suckers develop new roots within the first few years, but the parent root system remains alive and functioning for 40 to 50 years (Pregitzer and Friend 1996). Individual stems are relatively short lived. For example, most stems in the Colorado Front Range are less than 75 years old. Few stems reach 200 years of age. The wood of poplars lacks terpenoids and other compounds that resist decay, so the centers of large mature trees often are much rotted before they fall (Eckenwalder 1996).

Aspen has a broader range of environmental tolerances than most of its associated species and can grow in most mountain vegetation zones (Daubenmire 1943, Fowells 1965). In the intermountain region, however, aspen is confined to sites with moist soils with at least 38 cm of annual precipitation and cold winters with deep snows (Jones and DeByle 1985). Subsurface moisture from seeps, or other factors that concentrate water, characterize aspen sites. In western Wyoming, the modal elevation of aspen stands is about 2,000 m. The upper elevation limits may be determined by growing season length and the lower limits by evapotranspirational demands (Mueggler 1988). Aspen in the northern range of YNP occurs in small- to medium-sized stands growing on moist areas of the landscape. However, aspen stands in YNP are not as robust as stands in the mid-intermountain regions of Utah and Colorado (Mueggler 1988). In the Rocky Mountains, aspen is primarily a clonal species that reproduces almost exclusively by root sprouting and produces large stands composed of stems from one or a few genetic individuals (Barnes 1966, McDonough 1985, Tuskan et al. 1996). The individual clones respond differently to environmental conditions across the landscape. This is obvious in the fall when stands change color at different times and to different shades; differences may also be evident in responses to moisture stress and other environmental factors.

Reestablishment of aspen by seed is believed to have occurred infrequently in the Rocky Mountains since the last glaciation because climatic conditions have not been suitable for widespread germination and seedling establishment (Einspahr and Winston 1977, Cook 1983). However, rare episodes of seedling recruitment have occurred (Jelinski and Cheliak 1992), among them the widespread establishment of aspen seedlings in YNP after the 1988 fires (Kay 1993, Romme et al. 1997, Stevens et al. 1999). It is difficult to age aspen clones, as they are long-lived and may be much older than the oldest live canopy tree (Grant 1993). Existing large trees in an aspen clone represent "recruitment events" for tree-sized stems—that is, periods when conditions were suitable for seedlings or root sprouts to develop into tall trees.

Aspen occurs both as successional and as climax vegetation. Aspen root sprouts readily after disturbances such as fires that kill the overstory trees (Mueggler 1988, Romme et al. 2001). Fire enhances this recruitment, but any event, including aging, senescence, and death (e.g., by girdling) of the overstory trees, that reduces the apical dominance that typically suppresses root sprout development and growth may trigger extensive root sprouting. The extensive aspen stands throughout the central Rockies are believed to be partly a result of wildfires, and some investigators have suggested that the elimination of fire reduced regeneration of aspen (Loope and Gruell 1973, Houston 1982). Root sprouts tend to grow faster than seedlings, and thus aspen sprouts can easily outcompete other species, such as conifers, that must regenerate through seeds (Despain 1990). Aspen is not shade-tolerant, so if conifers start to grow under aspen stands, they may eventually shade out the aspen. In other locations, aspen can remain as the climax vegetation, continuing to regenerate through root sprouts when the individuals that compose the canopy become senescent (Despain 1990).

Aspen, like most *Populus* species, is preferentially browsed by ungulates when in leaf and during winter (Olmstead 1979). Elk eat the tips of aspen sprouts and the bark of mature trees, except where the smooth white bark has been replaced by thick, black, corky bark in response to prior injury. Aspen root sprouts and seedlings may be severely browsed by elk during winter or during spring and fall migrations between the summer and winter ranges. In many parts of YNP and the GYE, elk commonly browse nearly all root sprouts in aspen stands (Romme et al. 1995, Ripple and Larsen 2000a). Several investigators have suggested that excessive elk browsing is the major reason for the lack of regeneration of aspen stands (Krebill 1972; Beetle 1974, 1979; Kay 1990; Bartos et al. 1994). Indeed, ungulate browsing can increase mortality of aspen saplings and suckers reduce or eliminate root suckers (which prevents regeneration of large stems), and contribute to increased disease in larger trees (DeByle and Winokur 1985, Hinds 1985).

Aspen stands are scattered throughout the northern range where moisture conditions are suitable. Historical photographs suggest that aspen thickets occurred on the northern range during the early twentieth century (Houston 1982, Kay 1990, Meagher and Houston 1998). Kay (1990) and Keigley and Wagner (1998) show photographs of aspen stands in the 1800s with no browse line, whereas Houston (1982) describes a browse line for this period. In the late 1800s, there were approximately 6,000 ha of aspen on the northern winter range, whereas today aspen cover only about 2,000 ha, and many of them are in a shrub form (Renkin and Despain 1996). Because much of the northern winter range is sagebrush-grassland or wet bottomlands, locations where aspen can develop are limited. This scarcity of sites is common in semi-arid environments such as the lower elevations of the northern winter range. In the forested areas of the northern range, aspen represents a localized vegetation type, possibly relics of past fire regimes (Houston 1982).

Except for aspen clones in exclosures, most aspen stands on the northern range contain aging or senescent trees with a relatively dense understory of root sprouts mixed with a herbaceous ground cover. About 85% of the large aspen alive today on the northern range originated before 1920 (Romme et al. 1995, Ripple and Larsen 2000b). Romme et al. (1995) obtained increment cores from 15 aspen stands, which revealed a period of aspen regeneration in the 1870s and 1880s. Ripple and Larsen (2000b) developed a more comprehensive age structure for northern range aspen by obtaining 98 readable increment cores from 57 different stands. Results of this study revealed that 85% of the sampled aspen stems originated between 1871 and 1920, with 10% originating before 1871 and only 5% from 1921 to 1998. The relative paucity of older aspen (Romme et al. 1995, Kay 1997, Ripple and Larsen 2000b) is consistent with the relatively short lifespan of individual aspen stems. It is clear that very few of the stems became established in the twentieth century. However, quantifying the establishment of tree-sized aspen earlier in the nineteenth century is difficult because many trees have already died and some older trees are not sound enough for aging.

A historical data set in which the diameter at breast height (DBH) was recorded for aspen trees in 20 separate riparian stands in 1921 and 1922 (Warren 1926) has provided a valuable insight on northern-range aspen in the eighteenth and nineteenth centuries. Ripple and Larsen (2000b) used these data to infer historical age distribution. New increment cores were collected in

1998 from 30 northern range aspen and 19 aspen from the Gallatin National Forest to determine the relationship between aspen age and DBH. The resulting regression was then applied to the largest diameter aspen in each of Warren's (1926) stands. This analysis revealed that the aspen present in 1921 and 1922 had originated between 1750 and 1920.

Northern range aspen have produced root sprouts during recent decades, but root spouts have not been able to grow higher than the herbaceous layer. Sprouts are regularly browsed back by ungulates to the approximate depth of winter snowpack (Romme et al. 1995). Browsing intensity is high. For example, between 50% and 70% of aspen root sprouts in both burned and unburned stands on the northern range were browsed in 1990 and 1991 (Romme et al. 1995), and Ripple and Larsen (2000b) found that 89% of aspen stems in the northern range showed evidence of browsing in 1997. As the large, older aspen stems die, they are not replaced by new recruitment into the canopy, and shrubby aspen, grasses, and forbs may come to dominate (Kay 1990). However, root sprouts that have been protected from browsing within an exclosure have grown several meters, which suggests that climatic conditions are suitable for such growth. Removal of the exclosure exposes them to stem girdling by ungulates. These taller individuals then die, returning the clone to aged trees and herb-height root sprouts.

Multiaged or multiheight aspen clones are scarce in Yellowstone, especially in the northern parts of the park. In contrast, multiheight aspen clones exist at mid-elevations in areas within the GYE but outside YNP (e.g., Centennial Valley and Grand Teton National Park). Some areas of transitional range outside YNP may receive less browsing because migrating elk may not be present long enough or densities of ungulates in these areas may be lower than in YNP. Studies from Yellowstone and elsewhere in the Rocky Mountains provide insight into the regional variation in the factors that explain these differences, including fire, elk browsing, and other environmental drivers (e.g., DeByle and Winokur 1985, Romme et al. 1995, Baker et al. 1997, Kay 1997, White et al. 1998).

Factors Influencing Present Conditions of Aspen Communities

Precipitation

Rain and snow, by recharging subsurface soil moisture and protecting woody plants, affect the availability of winter forage and hence migration of ungulates. Together or separately, these may influence ungulate utilization of aspen root sprouts, shrub-aspen, and aspen bark. For example, in some reports on the northern winter range, long-term sprout height appeared to be a function of snow depth (Kittams 1952, Barmore 1980). Annual precipitation has fluctuated over the past century but without drying or wetting trends. During this time there has been no reduction in stream flows in general in YNP or in the Lamar River, as measured at a gage within the northern winter range. However, the 1930s and 1980s were drier decades than others in this century. Data from the 1980s have been used to infer a drying trend (YNP northern range report), but if there was a trend, it ended in the 1990s. Snow-water equivalent depth in April for two snow courses in the northern range shows a decline in the 1980s, a period that also was below normal for annual peak flows. The 1990s, however, had peak flows above normal, and heavy snow-packs in 1996 and 1997 produced back-to-back 100-year floods on the Yellowstone River.

The conclusions that can be drawn from the hydrological information are that there have been periods of above or below normal precipitation and snow cover during the past century, but there is no consistent pattern of declining precipitation or snow cover in the northern range. Consequently, snow through its influence on elk migration or protection of aspens, cannot be the one factor that produced changes in aspen stand conditions on the northern range. Precipitation as a source of subsurface moisture also has probably not changed sufficiently to affect aspen condition. The fact that aspen clones continue to produce prolific root sprouts and maintain shrub-aspen stands indicates that there probably is little moisture stress for these communities, although seedlings may experience greater moisture stress.

Balling et al. (1992b) examined the climate record for YNP (1890 to 1990) with regard to the relationship between climate and wildfire. Although they did not focus on snowpack, their analyses demonstrated a significant increase in summer temperatures during the century-long record along with a decline in January to June precipitation levels. Taken together, these changes suggest an increase in summer drought conditions in YNP during the past century. The period before the historical record coincides with the end of the Little Ice Age, an extended period of cooler global climate conditions, whereas the period of the Balling et al. (1992a) study ended before the wetter 1990s. Although these differences are not sufficient to explain the decline of aspen stands in YNP during the twentieth century, the climatic changes suggest the possibility of increased moisture stress as a contributing factor, although the twentieth century fluctuations in annual indices of drought stress fall within normal fluctuations over the past several centuries. However, in Yellowstone and throughout the Rocky Mountains, climatic change does not appear to be closely associated with aspen regeneration. In Rocky Mountain National Park, Baker et al. (1997) found little evidence for a relationship between aspen regeneration and climate, and aspen have not regenerated in Yellowstone during the climatically favorable conditions that have occurred since the 1920s (Ripple and Larsen 2000b). More important, their dendrochronological data show that aspen regeneration occurred during both favorable and unfavorable climatic conditions between 1871 and 1920 (Ripple and Larsen 2000b). Moreover, many aspen clones in the GYE where there is little winter use by ungulates show normal recruitment and expansion of clonal boundaries (committee member observations).

Fire

Fire is a major factor that enhances aspen recruitment (Loope and Gruell 1973, Brown and DeByle 1987). Typically, fire removes competing overstory conifers, triggering profuse aspen root sprouting and potentially producing extensive clonal stands. Reduction in fire frequency consequently may reduce the number and extent of aspen clones. Clones that do not have extensive successional understory development of conifers may continue to exist for many "generations" through root sprouting as the older trees senesce.

On the northern range, fire frequency has been reduced since the 1800s through fire suppression in YNP and adjacent national forests. Fire recurred every 20 to 25 years in northern YNP before the initiation of fire control in the late 1800s; from 1900 to 1988, there were almost no fires in Yellowstone's northern range (Houston 1973). Consequently, some of the aspen clones are becoming mixed aspen-conifer stands (Mueggler 1988). Others, however, remain as relatively pure aspen clones, especially in locations either some distance from conifer seed dispersal or on locations that may not be suitable for conifer recruitment. Examples of both pure aspen clones, some contained within exclosures, and aspen-conifer successional stands are found on the northern range. In both situations, there is often extensive aspen clonal root sprouting if the conifer canopy is not too dense. Thus, a reduction in fire frequency may reduce the expansion of existing aspen clones and also may prevent development of new clones through sexual recruitment. Fire suppression could be one factor contributing to the lack of recent recruitment of tree-sized stems. However, Hessl (2000) found no significant difference in aspen-sucker density in response to fire in the Gros Ventre watershed south of Yellowstone, and, even after the 1988 fires, aspen shoots have not grown enough to escape being eaten by elk in northern Yellowstone.

The 1988 Yellowstone fires burned approximately 22% of the northern range. Mature aspen stands that were burned in 1988 produced hundreds of thousands of root sprouts per hectare, but the 1988 fires resulted in a reduction in the number of aspen within shrub-aspen communities on the northern range (Kay and Wagner 1996). The density of shrub-aspen before fire (1986) was about 18,000 stems per ha, and shortly after the fire (1989) it was about 19,000 stems per ha, but by 1992 the number had dropped to about 10,000 stems per hectare. Extensive seeding of aspen in burned areas also occurred after the 1988 fires (Kay 1993, Romme et al. 1997). These seedling aspen are genetically diverse but are not elongating rapidly in YNP (Stevens et al. 1999). Browsing intensity on seedling aspen remains high, and even unbrowsed individuals attained maximum heights of only about 1.5 m by 1999 (M.G. Turner, personal observation). Thus, despite extensive root sprouting and seedling establishment after the 1988 fires, regeneration of large aspen is not yet occurring.

The accumulation of postfire coarse woody debris may, however, favor aspen regeneration. As conifers killed by fire fall, extremely dense piles of coarse woody debris may accumulate and provide refugia from ungulate browsing. After the 1988 fires, seedling aspen growing in areas with a lot of fallen conifers had not been browsed by ungulates since 1993 (Turner and Romme 1994). In burned areas on the northern range, Ripple and Larsen (2001) found that aspen suckers protected by fallen conifer barriers were, on average, twice the height of unprotected suckers.

Ungulate Use

Dendrochronological studies in the GYE and elsewhere in the Rocky Mountain region suggest a strong relationship between elk density and regeneration of aspen cohorts—stands containing a predominant size class or distinct patches of different size classes that are spatially segregated within a clone (Baker et al. 1997). In Rocky Mountain National Park, Baker et al. (1997) found only a weak correspondence between regeneration of aspen cohorts since 1875 and climatic and hydrologic fluctuations; aspen regenerated in both cool and warm periods and during periods with less than average precipitation. However, aspen regeneration seemed to be related to fluctuations in elk density, with cohorts regenerating when elk density was probably low and exhibiting no regeneration when elk density was high (Baker et al. 1997). Notably, aspen regeneration appeared to have been sporadic when elk density was lower than today (e.g., late 1800s to early 1900s), although older cohorts may have already senesced. The complete absence of recruitment of aspen trees since the late 1970s is anomalous in their record. Based on mortality levels of suckers and established trees, sucker density, and the reduced height and extensive branching of browsed suckers, Baker et al. (1997) concluded that the aspen population in the elk winter range in Rocky Mountain National Park is declining, largely in response to elk browsing. In the Gros Ventre watershed, located south of Yellowstone and managed by the Bridger-Teton National Forest, Hessl (2000) examined aspen regeneration since 1850 in relation to fire, elk browsing, climatic variation, and land management regimes. Regeneration of aspen cohorts in the Gros Ventre has been episodic and largely corresponds to periods when elk density was lower (e.g., late 1800s and 1940s and 1950s) or browsing pressure on native vegetation was reduced (1970s) because of artificial winter feeding in the National Elk Refuge located near Jackson, Wyoming. Current patterns of aspen regeneration appear to be related to human-induced gradients of elk use, with browsing intensity highest near the winter feeding grounds and declining with distance from the feeding grounds.

Ripple and Larsen (2000a) compared the percentage of aspen stems browsed in stands located in the northern range and in two locations in the nearby Gallatin National Forest: the Sunlight Basin, close to the northern range, and Clarks Fork Basin, which does not receive heavy ungulate use in the winter. Their initial data revealed that the highest browsing intensity occurred on the northern range (89.6% of 1,100 observed ramets were browsed), but browsing intensity was also high in the Sunlight (85.6% of 611 ramets) and Clarks Fork (76.7% of 322 ramets) Basins. However, the Clarks Fork site contained a more even distribution of aspen in a wide range of size classes compared with the northern range, in which all aspen stands contained individuals of >20 cm DBH and none of <11 cm DBH. In the Sunlight Basin, 90% of the stands contained aspen of >11 cm DBH, but 11.5% of the stands contained individuals of 1 to 5 cm DBH, and 27.8% of the stands had individuals of 6 to 10 cm DBH. Ripple and Larsen are also using aerial photos to examine changes through time in aspen stands in northern Yellowstone and the Gallatin and Shoshone National Forests, but the results of these analyses are not available.

There is indisputable evidence that ungulates, primarily elk, are browsing aspen intensively on the northern range and likely have done so throughout the twentieth century. Ungulates are a major factor contributing to the current absence of recruitment of tree-sized aspen in YNP (Kay 1993, Romme et al. 1995, Ripple and Larsen 2000b) and in other locations in the Rocky Mountains (Baker et al. 1997, Hessl 2000). What is not known, however, is how ungulate numbers and patterns of habitat use before park establishment might have influenced the spatial and temporal dynamics of herbivory on aspen. This perspective is important because an understanding of the dynamic interaction between elk and aspen would help us interpret the current condition.

There is controversy about the numbers, or even the presence, of ungulates in the northern range in the pre-Columbian period. Evidence supports the presence of ungulates (primarily elk) in the northern range when YNP was established (Schullery and Whittlesey 1992). However, available evidence does not suggest that extensive herds of ungulates wintered in what is now the northern winter range in the period immediately before establishment of YNP (see Appendix A). Certainly, photographic evidence presented by Kay (1990) and Keigley and Wagner (1998) suggests that there was little woody plant utilization by ungulates in the mid- to late 1800s. This could mean that (1) no or few ungulates were there any time of year (Kay's position), or (2) that no or few ungulates used the northern range when forage was scarce and woody plants were one of the few food sources available because herds had migrated to lower elevations, conditions expected during severe winters (Keigley and Wagner's position).

It is unlikely that woody plants, including aspen stands with little highlining and extensive tall willow communities, could have existed if large numbers of ungulates used the northern range in winter in the late 1800s. Data collected by Barmore (1965, presented by Renkin and Despain 1996) when the elk herd was reduced in the late 1950s and early 1960s show that levels of elk use as low as 10 elk-use days per acre resulted in over 75% utilization of aspen sprout leaders, and elk-use days no greater than 25 per acre resulted in 100% utilization. Consequently, there appears to be a change in the ungulate use pattern of the northern range over the past century and a half, with more animals staying in the park's northern range in the winter. This change may be due to several factors, none of which has been thoroughly tested. These include increased human activities and development within the original winter range and winter migration routes as well as development of a regular hunting season that created hunting pressures on the YNP northern border. Cole (1969) used these factors to account for altered migration patterns and forage use of elk in southern YNP-Grand Teton National Park elk herds.

Chemical composition, especially secondary compounds, may influence herbivory and the survival of aspen. Some authors have suggested that willows and aspen in YNP have inadequate secondary compounds to defend against elk browsing (e.g., Singer et al. 1994), but the chemical composition of aspen and its relationship to browsing have not been studied in Yellowstone. The dominant secondary metabolites of aspen are phenolic products of the shikimic acid pathway. These include condensed tannins, phenolic glycosides (the salicylates salicin, salicortin, tremuloidin, and tremulacin), and coniferyl benzoate (Palo 1984, Lindroth and Hwang 1996). Tannins and phenolic glycosides occur in leaf, stem, and root tissues, whereas coniferyl benzoate occurs only in flower buds. The influence of tannins and phenolic glycosides on feeding by insect herbivores, particularly Lepidoptera, has received considerable investigation. Tannins are ineffective defenses, whereas phenolic glycosides are effective at moderate to high concentrations against many aspen-feeding insects (Bryant et al. 1987, Hemming and Lindroth 1995, Lindroth and Hwang 1996, Hwang and Lindroth 1998). Coniferyl benzoate provides protection against herbivorous birds such as grouse (Jakubas et al. 1989). However, little is known about the influence of aspen secondary metabolites on the foraging behavior of mammals, particularly ungulates. Phenolic glycosides and their derivatives, but not tannins, deter feeding by hares (Tahvanainen et al. 1985, Reichardt et al. 1990). Aspen tannins may be similarly ineffective against browsing ungulates, a prediction consistent with the production of tannin-binding salivary proteins in such animals (Austin et al. 1989, Hagerman and Robbins 1993). Jelinski and Fisher (1991) found significant variation in secondary compounds among aspen clones but concluded that these concentrations did not appear to inhibit palatability or digestibility for members of the deer family. Erwin et al. (2001) concluded that for aspen seedlings and clones in Yellowstone, foliar phenolic glycosides and tannins were not active defenses induced in response to browsing by large mammals.

Several exclosures in the northern range and the Gallatin range (northwestern YNP) demonstrate that, with protection from ungulates, aspen can still grow tall and root sprouts can produce multiple age classes of above-ground stems. Intensive ungulate browsing on aspen may result from factors that concentrate the animals near the aspen clones, and/or from the ungulate herds being sufficiently large that nearly all aspen is utilized. If ungulates are "prevented" from migrating to portions of their historical winter range, the density of ungulates on the remaining available winter range will be unnaturally high. Hunting and human developments (e.g., throughout Paradise Valley, north of YNP) can prevent long-distance migration. Truncation of migration routes may also hasten the spring return of ungulates to the summer range at times when herbaceous plant biomass is still low and woody forage must be consumed.

Interrelated Changes

Aspen communities no longer produce root sprouts that grow into tall trees in the northern range. Root sprouting continues, which indicates adequate local abiotic conditions for this form of recruitment and suggests that other factors must be contributing to prevent the development of tall aspen clones. Fire suppression has prevented reduction of competition and enhancement of conditions for tall growth. However, this cannot explain the lack of regeneration, particularly given that the 1988 fires resulted in profuse root sprouting that was intensively browsed. Ungulate browsing clearly contributes to the lack of regeneration, although the long-term dynamics of aspen in northern YNP are not well understood. Romme et al. (1995) hypothesized that during the late 1800s, when they thought the last recruitment of tree-sized aspen occurred, might have been characterized by several co-occurring factors, including intensive market hunting for elk, more frequent fires, generally moist growing conditions, and reduction in beaver populations, which all contributed to reduced browsing pressure. However, recent work by Ripple and Larsen (2000b) indicates that aspen regeneration did occur as recently as 1921. In interpreting these patterns, Ripple and Larsen (2000b) hypothesized that the disruption of natural predator-prey relationships may have contributed to the observed differences in aspen regeneration. Furthermore, they suggest that the reestablishment of wolves in 1995 may benefit aspen in the long term. In addition to reducing elk population size, wolves may also influence ungulate movement and browsing patterns. In Jasper National Park, White et al. (1998) reported that a new cohort of aspen sprouts regenerated into trees ranging from 3 to 5 m tall after wolves were reintroduced. Top-down control on ungulate herbivory has also been suggested in Isle Royale National Park, a system with no long-range ungulate migration (McLaren and Peterson 1994).

Postfire coarse woody debris (i.e., fallen trees) also can protect aspen from ungulate browsing and has contributed to greater elongation of aspen suckers in burned stands on the northern range (Ripple and Larsen 2001). Ripple and Larsen (2001) have also suggested that park managers might create experimental "jackstraw piles" of dead conifers to create barriers to browsing and hence facilitate aspen regeneration.

Conclusion

All data sources indicate that the abundance of large aspen in northern Yellowstone has declined during the twentieth century, and most indicate that this decline is due primarily to ungulate browsing. Although climate and fire may influence aspen dynamics, these factors have not been the main drivers for the changes in aspen observed during the recent century. Rather, the ungulate population, both in size and behavior, appears to be most strongly correlated with aspen regeneration in the GYE. Large aspen trees are likely to regenerate on the northern range under current conditions only if they are protected from ungulates, either through physical barriers (which may develop naturally with postfire coarse woody debris or which may be provided artificially) or through behavioral changes, as might be induced by the presence of predators such as wolves, unless much more extensive lower-elevation winter range becomes available.

RIPARIAN COMMUNITIES OF THE NORTHERN RANGE

Riparian ecosystems are the transition from stream to upland. They occupy a very small part of the landscape, often less than 1%, yet they play an important role in stream dynamics, wildlife ecology, and biodiversity of the region (Naiman et al. 1993, Naiman and Decamps 1997, Patten 1998). In most cases riparian ecosystems occur on alluvial sediment deposits where river and alluvial groundwater supplements water available from precipitation (Gregory et al. 1991). Riparian ecosystems may also be found near springs and seeps where the groundwater surfaces to create wet areas or surface flows. Riparian ecosystems are frequently disturbed by periodic flooding and thus are in a continual process of succession (Malanson 1993). They also exhibit a high level of resiliency after termination or removal of disturbances or stressors—for example, after floods, after the return of normal stream flows, or after removal of grazing (Stromberg et al. 1997; Kaufmann et al. 2000; Patten 1998, 2000).

Most western North American riparian vegetation communities are a result of recruitment and survival in response to seasonal hydrological events, variation in groundwater depth, and flood-generated gravel bars. For example, most western cottonwood species recruit along streams on bare, moist surfaces during the decline of spring high flows (Friedman et al. 1995, Scott et al.

1997, Stromberg et al. 1997). Survival of these trees depends on maintenance of a high water table in the floodplain and avoidance of scouring by floods and ice flows. Mortality, or inability to survive after recruitment may result if the water table drops below that tolerated by young or maturing plants (Rood and Mahoney 1995).

Willows are a common woody component of many western riparian communities. Although their recruitment may also coincide with spring floods, availability of damp soils may be sufficient for them to establish and survive. Factors that help maintain an elevated water table enhance willow community growth and expansion. For example, beaver activity may elevate the water table and create suitable sites for willow community expansion, whereas stream incising may lower the water table and cause stress to willows.

In the Rockies, valley geomorphology directly influences the extent and type of riparian communities (Patten 1998). Streams that flow through broad valleys with low gradients may be lined by both woody and herbaceous riparian vegetation. If the water table is shallow, wetland herbaceous plants (e.g., sedges and wetland grasses) may extend for some distance from the river. These areas often are devoid of woody species because the herbaceous plants may prevent establishment of willows or cottonwoods. Willows and sometimes cottonwoods may occur near the stream, where floods enhance their recruitment. Once established, these species may spread asexually and expand within the floodplain, often occurring away from the stream as it migrates across the floodplain.

Riparian communities in the northern range have developed in response to many of the conditions discussed above. Both low-gradient and relatively steep valleys exist in the northern range. In the lower reaches of the northern range, the Yellowstone and Gardner Rivers cut through canyons, providing only narrow areas for cottonwood stands and shrub-willow communities. Twenty-four species of willow are found in the northern range, and species composition changes, in part, with elevation as well as geomorphic setting (YNP 1997). For example, false mountain willow (*Salix pseudomonticola*) is more common at lower elevations; Drummond, Farr, and Barclay willows (*S. drummondiana, S. farrii, S. barcleyi*) are more common at higher elevations; wolf willow (*S. wolfii*) is common on broad floodplains; and sandbar willow (*S. exigua*) is common on sandy, exposed-stream meander lobes. Other woody species, such as shrubby cinquefoil (*Potentilla fruticosa*) and water birch (*Betula glandulosa*), are also found in the riparian zone (Singer 1996).

Reaches of the Lamar River and Soda Butte and Cache Creeks at mid-elevation in the northern range extend across low-gradient floodplains where willow and wetland herbaceous communities are common. Narrow leaf cottonwood (*Populus augustifolia*) occurs in limited stands near the channel on present and past river meander lobes (point bars) (Keigley 1997b). At higher elevations these rivers flow through both narrow- and steep-gradient valleys and shallow low-gradient floodplains. Cottonwoods drop out and willow and herbaceous communities dominate on low-gradient reaches. Conifers become co-dominants along the streams as the range changes from winter/summer to only summer ungulate range (the upper elevations of the northern range). Throughout the northern range are several seeps and springs that also support deep-rooted communities, primarily willow. These are not truly riparian communities because they do not occur along the edge of streams or lakes, but the shallow water table that creates the spring or seep also creates suitable habitat for riparian and wetland species. Also, in some areas aspen is a riparian species. Warren (1926) found aspen along streams and ponds, especially where there was extensive beaver activity, but most of those stands are gone. Aspen can still be found in riparian conditions but in limited situations, especially along smaller streams. Most aspen stands in the northern range are in moist upland areas.

Riparian communities in the northern range exist, or have existed, because conditions were suitable (1) for recruitment of the characteristic riparian species, (2) for maintenance of established species because the alluvial water table was sufficiently shallow to maintain plant growth and survival, and (3) because factors, such as loss of adequate groundwater availability, changes in surface hydrology, plant community modification through browsing and grazing activities, and other human modifications of the landscape exerted only moderate pressures. If any of these conditions—which helped create the riparian communities—change through natural or anthropogenic actions, the riparian community is stressed and may degrade or be lost from places on the northern range where they formerly occurred.

The riparian ecosystems along the northern borders of YNP and adjacent areas of ungulate winter use, including the northern range and the Gallatin River valley in the northwestern corner of the park, show evidence of degradation in response to stressors (Patten 1968, Singer et al. 1994, Keigley 1997b, YNP 1997). In these areas, woody riparian vegetation has many dead branches, often with new growth emanating from low on the plant. In the upper Gallatin River drainage there are no cottonwoods, and so evidence of stress exists only in willows, whereas in the lower and mid-elevations of the northern range both willows and cottonwoods show signs of stress. In some cases, where factors that maintain riparian communities have changed substantially during the past 100 years or more, riparian communities are absent. In other situations, woody riparian vegetation appears to be stunted or maintained in a short-stature growth form. This is true of both cottonwoods and willows. Evidence of periodically successful recruitment of young cottonwoods and of the existence of cottonwood canopy layers representing several age classes or recruitment events is almost lacking along most of the streams in the northern range. Mature and overmature cottonwoods occur in several locations, and hedged cottonwoods with large stems, which clearly are old, are sometimes found in these mature cottonwood galleries (Keigley 1997b). Cottonwood seedlings are found on good recruitment sites some years, but these seedlings currently are not surviving to become mature trees. Although narrow-leaf cottonwood can root-sprout and this is occurring along the streams in the northern range, root sprouts also do not appear to be reaching maturity or growing tall enough to contribute to a mid-height or tall canopy within the cottonwood galleries.

On broad, low-gradient floodplains, extensive willow communities still exist, but they are generally overtopped by the associated herbaceous vegetation. In other places where willow occurs, such as at seeps or along steeper gradient streams, willows also appear to be prevented from growing to normal stature for the species. Willow species differ in their normal height growth. For example, wolf willow generally is of low stature (1 to 1.5 m), whereas Geyer, Booth, and Bebb willow generally attain heights of several meters. Many willow species can reproduce asexually, and spread along the streams where moisture conditions are favorable. Recruitment of shrubby willow species is not tied as closely to hydrological events as that of cottonwood. Seed dispersal may not depend strongly on spring floods, and recruitment may occur on any available moist substrate that is accessible to the seeds. Willows normally require and are more tolerant of wetter soils than cottonwood. Consequently, their establishment in near-wetland conditions close to streams, instead of on more elevated locations (0.5 to 1.5 m above baseflow) where cottonwoods grow, shows they could grow along a river or on any wet soil with persistent alluvial groundwater regardless of surface events such as spring floods. Willows are also less tolerant of drought conditions and lower water tables than cottonwoods (Stromberg et al. 1996).

Factors Influencing Present Conditions of Riparian Communities

Hydrology

Several hydrological factors influence riparian vegetation recruitment, growth, and maintenance. We discuss these factors separately, as changes in any one may significantly influence riparian communities.

Precipitation

Precipitation influences riparian ecosystems through its influence on runoff, stream flow, and groundwater recharge. Variability in precipitation for the past several decades falls within normal variation over the period of record, including expected periods of drought and above-normal precipitation. If precipitation were the cause of changes in riparian vegetation in the northern range, stream flow magnitudes should be outside the normal range. An analysis of peak and annual flows in the Yellowstone River near Corwin Springs shows that both peak flows and annual discharge volumes were generally below average in the earlier part of the twentieth century (Figures 3-1 and 3-2). Peak flows are analyzed in addition to annual discharge because they often closely represent snowpack conditions, the conditions that recharge and maintain elevated water tables in the watershed. Corwin Springs was used because it is the first U.S. Geological Survey (USGS) gage outside YNP on the Yellowstone River. The Yellowstone River drains most of the watershed of the northern range in addition to southern watersheds in Yellowstone, and data from the Corwin gage integrate most of the watershed output. Discharge patterns at Corwin Springs also correlate well with discharge patterns of the Lamar River, which primarily drains the northern range (Mowry 1998). Precipitation gages and snow measurements are spotty, and precipitation in mountainous terrain tends to be quite heterogeneous. Thus, an integrating measurement is more useful.

YNP used Yellowstone's northern range data to argue that there was a decline in precipitation and stream flows from 1982 to 1994 caused by drying conditions that stressed the range, forest, and riparian ecosystems (YNP 1997). However, this is too short a period to demonstrate unusual variation. Peak flows and annual discharge averaged below normal during the 1980s, but

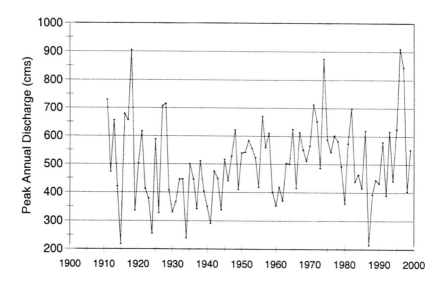

FIGURE 3-1 Annual peak stream flow in cubic meters per second (cms) at Corwin Springs on the Yellowstone River a few miles downstream from Gardiner, MT. USGS gage 06191500.

willows declined (i.e., reduced stature and loss) during the earlier decades of the twentieth century (Smith et al. 1915, Warren 1926). Therefore, it is necessary to use hydrological data from most of the century to detect whether hydrological changes have caused willow decline.

Increased precipitation at high elevations has been used to explain the vigor of higher-elevation willow communities (YNP 1997); however, willows occur along stream courses where they have access to groundwater and soils wetted from the stream or from capillary rise of water from the water table (Dawson and Ehleringer 1991, Busch et al. 1992, Flanagan et al. 1992). Consequently, the amount of local precipitation has very little influence on willow growth and survival, except when it is in the form of a snow cover (see below).

Snow Accumulation

Annual snow accumulation during the twentieth century was within normal variability for the period of record based on spring runoff data (Farnes 1998).

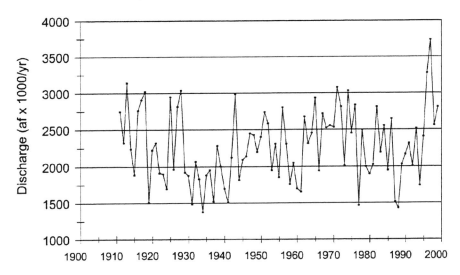

FIGURE 3-2 Annual discharge volumes (thousands of acre-feet per year) at Corwin Springs on the Yellowstone River a few miles downstream from Gardiner, MT. USGS gage 06191500.

There has been considerable variation in the past few years, with 1996 and 1997 having high accumulations, and 1998, 2000, and 2001 having accumulations well below the mean. Some evidence shows that runoff from snow melt peaked earlier, by three days on average, after the 1988 fires (Farnes 1998) and that peak flows may have been higher, but peak discharge data do not support the latter conclusion. Also, runoff appears to have returned to normal for the snowfall amounts within a decade after the fires. Snow accumulation may play a more important role in riparian vegetation structure through protection of riparian shrubs as pointed out by Singer (1996). Not only does deep snow accumulation around riparian woody plants reduce browsing use, it also may prevent use of higher-elevation riparian vegetation by ungulates. Conversely deep snow can cause greater use of riparian vegetation by browsers if herbaceous plants are buried and woody stems are the only forage exposed. This is especially true when a thaw-freeze cycle creates an impenetrable ice layer over herbaceous vegetation. The present condition of riparian vegetation in the northern range may be caused by heavy utilization when it is one of the few sources of forage in deep snow years or years with significant thaw-freeze events.

Stream Discharge (Hydrographs)

Streams often recharge adjacent groundwater, although in some cases groundwater movement may be toward the stream rather than away from it. Consequently, reduction in stream flows combined with reduced groundwater movement from uplands when the precipitation is below normal may reduce groundwater availability to riparian vegetation. However, the USGS daily historical hydrological data for rivers flowing from the northern range (e.g., Yellowstone at Corwin Springs) do not indicate any unusual reduction or increase in stream flows over the past century beyond normal variation. Consequently, surface hydrological changes do not appear to be sufficient to have directly affected riparian availability of groundwater, although if some streams incised their channels during the past several decades, that could have lowered the surface flow enough to lower the alluvial water table to levels that could stress riparian vegetation.

Spring floods can scour channel margins and, through overbank flows, scour and flood the adjacent floodplain to produce suitable sites for recruitment. High spring flows that could enhance recruitment of riparian plants may occur only every 5 or 10 years, resulting in spaced age-classes of the woody riparian species, which is especially evident in large woody species such as cottonwood. The lack of evidence of regular recruitment events in the northern range indicates that perhaps there have been fewer spring flood events during the past few decades. However, USGS peak flow data for rivers coming from the northern range watershed show periodic high or flood flows, sufficiently high to produce a recruitment event and probably high enough to have seedling establishment above either ice scour or smaller scouring flood events. Therefore, the lack of periodic recruitment by riparian plants in the northern range cannot be attributed to absence of flood events.

Groundwater (Alluvial Water Table): Streams, Seeps, and Beavers

Groundwater monitoring data obtained by the committee show that groundwater in the floodplains, where most of the riparian vegetation in the northern range occurs, has not changed sufficiently in the past several decades to cause stress on riparian vegetation. Data from one well north of Gardiner showed water table decline in the 1980s (YNP 1997), but the 1980s were a dry decade and domestic use of groundwater in the area of the well probably increased as well. Groundwater monitoring wells within the northern range are needed if seasonal and annual fluctuations in groundwater levels are to be determined.

The existence of riparian vegetation, especially willow, along most of the rivers and on the floodplains in the northern range indicates that the water table is still high enough to maintain these communities. The occurrence of seeps and springs is additional evidence that the water table is near enough to the surface to maintain riparian vegetation. Groundwater change appears to have been an important factor in those few areas where riparian vegetation once occurred but is no longer present because the beaver ponds were abandoned.

Stem xylem water potential can indicate water stress in woody plants. Measurements of xylem water potential of short (browsing suppressed) and tall stature willows near exclosures showed water potential averages of -1.17 ± 0.68 and -2.4 ± 0.85 bars, respectively. Comparisons between short and intermediate stature willows were -1.71 ± 0.68 and -2.86 ± 1.35 bars, respectively (Singer et al. 1994). Thus, tall and intermediate willows with greater canopies had greater water stress than short willows, but these values are much less than wilting-point values (i.e., -15 bars) for mesic plants. Loss of willows during the 1988 drought period may be partly explained by water stress (Singer et al. 1994) but may be better explained by possible water table declines and warm, dry winds that occurred during that period.

Beavers, whose activity currently helps maintain riparian vegetation along many streams in the GYE, are currently rare or absent in most of Yellowstone's northern range. The reasons for beaver decline and the consequences for willows have recently been controversial (YNP 1997; Singer et al. 1998, 2000; Keigley 2000). An early theory attributed beaver decline to reductions due to ungulate browsing of woody plants (willow and aspen), which are beaver food and building material (Bailey 1930, Wright and Thompson 1935, Jonas 1955). In the absence of beavers, water tables in areas elevated by beaver dams may decline so that riparian vegetation cannot survive. Singer et al. (1998, 2000) argued that this process is exemplified by the healthy stands of willow at Willow Meadows (a location in the transition zone from winter to summer northern range use) that occur on a broad floodplain where the water table is less than a meter deep are maintained by beavers.

Geomorphology: Stream Banks and Channels

Many of the streams in the northern range on the broader floodplains now produce braided channels where formerly they produced single or multiple meandering channels (Meyer et al. 1995). The braided channels are dynamic and have moved considerably over the past several decades (Mowry 1998). Changes in the width of Soda Butte Creek measured when it was full to the banks near the upper edge of the winter range were related to hydrological history and perhaps vegetation cover (Mowry 1998). Locations with tall willows and little winter ungulate use remained stable (i.e., no statistically significant changes) during a relatively constant hydrological period from 1954 to 1987, whereas channels with grass-covered banks and reaches with little willow bank stability narrowed over that time. That demonstrates, for these sections of Soda Butte Creek upstream of most wintering ungulates, that vigorous woody riparian vegetation tends to maintain bank stability under most hydrological conditions whereas non-woody banks tend to aggrade, a process explained by Lyons et al. (2000) from research on Great Plains rivers.

However, the floods of spring 1996 (one of two back-to-back hundred-year floods on the Yellowstone River and its tributaries) altered the willow-stabilized banks more than banks stabilized by grass or low willows. That happened because woody riparian species (e.g., willows) stabilize banks only to a flood-discharge threshold, beyond which shear stress scours and uproots the plants, so that banks collapse (e.g., Stromberg et al. 1997). Stream banks that are stabilized only by vegetation that offer low resistance to flowing water (e.g., grasses and low willows), may withstand high-velocity floods better than banks with woody plants. Also, if one high-flood year is followed by another, as was the case in 1996 and 1997 in Yellowstone, the instability created by the first year may allow the subsequent flood to alter the channel morphology more than it would if it had occurred alone. For example, the 1997 hundred-year flood on the main stem of the upper Yellowstone River downstream from YNP removed many hectares of mature cottonwood trees that had been made unstable by the 1996 flood (committee observation).

The sandy deposits in several of the rivers of the northern range, primarily the Lamar River, are thought by some observers to be a product of increased erosion along the river channels (R. Beschta, Oregon State University, comments to committee, 1999) and high sediment supply from steep erodible terrain and tributary streams (Rosgen 1993). Rosgen suggested that poor riparian conditions due to excess browsing cause unstable channel conditions and increased sediments in the Lamar River valley. In addition, Chadde and Kay (1991) claimed that the Lamar River valley channels have incised, perhaps as much as several meters, because of heavy ungulate use. If incision has occurred, the shallow riparian water table needed to sustain riparian communities and vigorous growth response to browsing will not be maintained. For example, willows growing along an apparent water table gradient away from the

Gallatin river have decreased growth and recovery to browsing with distance from the river (Patten 1968).

Floodplains with shallower alluvial deposits, or steeper low-elevation valleys, tend to have bedrock conditions nearer the surface that prevent decline of the water table even if stream flows are reduced. In such places, willow vigor is maintained even when browsing occurs.

Ungulate Use

Several studies have assessed the response of willows to protection from browsing (Kay 1990, 1994; Chadde and Kay 1991; Singer et al. 1994, 1998). Some of these studies have used measurements that show that frequency and occurrence of willows inside and outside exclosures do not differ significantly (YNP 1997). These measurements count the number of individuals present, regardless of their size or vigor. Therefore they cannot detect changes in size, cover, or productivity—key vegetation components. Other studies have measured willow cover (Kay 1990, 1994; Singer et al. 1994; Singer 1996), which is a better metric for determining shrub vigor because it measures actual areal cover of the canopy. Singer (1996) showed that, during the three decades after construction of upland exclosures, total canopy areas for several species of willow inside the exclosure were 200% to 600% greater than those outside. Also, annual biomass production was 5 to 10 times greater inside the exclosure than outside (Kay 1994, Kay and Chadde 1992). During the period between measurements of willow canopy cover in the 1950s and 1960s and measurements in the 1980s, cover increased many fold inside and hardly changed outside exclosures (Kay 1994). Data from exclosures showed that only willows protected from browsing reproduced (Kay 1994). No seed-producing catkins were found outside, but they averaged over 300,000 per m² inside. Ungulate browsing also removes pollen-producing catkins. Near Geode Creek, no catkins were found on willows below the browse height, but many were found above the browse height (Kay 1994). This reduction in reproduction of willows in the northern range may account for the reduction in willow pollen in recent lake sediments (Barnosky et al. 1988).

Although exclosure studies show that browsing has a dramatic effect on willows, exclosures represent the extreme of no browsing—rare in ecosystems that have browsers, as YNP does. Thus, the best way to determine how ungulate populations within and outside YNP cause changes in willow communities is to compare natural willow communities inside and outside YNP but still within the northern range. Unfortunately, this type of study has only been done for aspen (Kay 1990), although several recent publications comparing early and recent park photos show that tall-willow communities have declined in the northern range (Kay 1990, YNP 1997, Keigley and Wagner 1998, Meagher and Houston 1998). Explanations for the differences between early and recent photos include excessive browsing by ungulates (Kay 1990, Keigley and Wagner 1998) and climate change, especially during the 1930s drought (YNP 1997, Meagher and Houston 1998). Groundwater conditions, which have not been measured in any of the studies of ungulate utilization of riparian shrubs, should be measured at all sites.

The lack of cottonwood recruitment may be due to ungulate browsing. When ungulate numbers were reduced in the northern range in the 1950s and 1960s, cottonwoods were temporarily "released," put on new shoots, and grew taller than during the preceding or following decades (Keigley 1998).

Concentrations of defensive or secondary chemicals in riparian vegetation may influence the magnitude of ungulate use. Several studies were designed to determine whether low-stature willows have lower concentrations of defensive chemicals and thus are more palatable and more heavily used by herbivores than taller willows (Singer et al. 1994). Extremely short stature is not a natural condition for most willows. Consequently, very short shrubs must be a result of browsing. A controversy has developed about whether the level and composition of certain secondary chemicals is a result of short stature or a response to browsing. This controversy has resulted in a series of commentary papers following Singer et al. (1994) (Singer and Cates 1995, Wagner et al. 1995).

Studies on secondary chemicals in willows generally show that low-elevation, short willows tend not to produce the concentrations of defensive secondary chemicals (e.g., tannins) found in higher-elevation and tall willows. This supports the argument that the tall willows have greater defenses against ungulate browsing and thus remain tall because ungulates avoid them. However, many of the tall willow communities are at higher elevations where snow cover or other conditions prevent heavy winter ungulate use. Thus, tall willows occur in areas with few wintering elk. Also short willows might not produce as high a concentration of secondary chemicals as tall willows because growth after browsing, the only type of shoot these short willows produce, tends not to produce as many of these chemicals. One question that is not answered in the many studies dealing with secondary chemical defense is whether the concentrations of secondary chemicals found in tall-stature willows are sufficient to prevent browsing even by ungulates that are starving. Starving ungulates will eat almost any woody plant that is available if other food such as grass and forbs are buried and unavailable. Browsed highlines on lodgepole pine, a low-quality food, are an example of this type of ungulate use. The evidence that lower levels of defensive chemicals increases utilization of willows or other riparian shrubs by browsers is insufficient to explain the decline in stature or loss of willows and riparian vegetation in the northern range during this century.

Conclusions

During the first six decades of the twentieth century the NPS was concerned that ungulates were reducing the riparian communities; however, since then, factors other than ungulate browsing, such as climate change, have been hypothesized to explain the loss or reduction in stature of the riparian communities of the northern range (Houston 1982, YNP 1997). The committee concludes that some riparian losses may be due to changed hydrological conditions in addition to responses of vegetation to ungulate use. Flooding events and water tables are still suitable for recruitment of the dominant riparian vegetation. However, channel incision might have lowered the water table, reducing the ability of willows to recover from browsing. The increase in ungulate browsing over the past century, as evidenced by hedged willows and lost willow stands throughout lower elevations of the northern range, has caused most of the reduced willow cover and lowered willow reproduction this area. Ungulate use also appears to be the primary factor preventing cottonwoods from recruiting, because seedlings do appear on good seedbed sites along the rivers of the northern range after appropriate hydrological events, but they fail to survive.

WETLAND VEGETATION OF THE NORTHERN RANGE

Wetlands are a prime source of biodiversity in YNP (Elliot and Hektner 2000). Several types of wetlands, such as natural depressions, beaver dam wetlands, thermal wetlands, and wetlands along rivers and creeks (i.e., riparian wetlands), are found on the northern range. Chadde et al. (1988) defined 62 wetland communities in the northern range. These included wetlands dominated by trees (e.g., spruce [*Picea*] and aspen), willows, shrubs (e.g., cinquefoil [*Potentilla fruticosa*] and silver sagebrush [*Artemisia cana*]), and graminoids (e.g., grasses and sedges). Most wetlands in YNP have been mapped under the U.S. Fish and Wildlife Service's National Wetlands Inventory mapping program. Although wetlands are important across YNP, they are hardly mentioned in the park's documents on the northern range (YNP 1997) or in a vegetation description of YNP (Despain 1990).

There have been limited studies of wetlands in the northern range. Houston (1982) sampled five highly productive wetland meadow sites with humic soil types dominated by sedges. Relatively few species were sampled (1 to 13 species on the five sampled sites). The sites were dominated by sedges and rushes and graded to species characteristic of mesic grasslands.

One wetland type was identified as tufted hairgrass-sedge type, described by Mueggler and Stewart (1980). This habitat type dominated by *Deschampsia caespitosa,* is found on high-elevation valley bottoms between 2,000 and 3,000 m (Mueggler and Stewart 1980). The soils are deep and poorly drained, with water standing on the soil surface at least part of the growing season. *D. caespitosa* is the dominant grass but sedges (*Carex* spp.) are always present. Other grasses include *Danthonia intermedia, Phleum alpinum,* and *Agrostis* and *Juncus* spp. Forb species present include *Potentilla gracilis, Polygonum bistortoides,* and *Antennaria corymbosa.* This is a highly productive habitat type with annual production reaching 2,900 kg/ha.

Other wetlands were on alkaline soils formerly dominated by alkali grass and meadows cut for hay, now dominated by introduced timothy grass. All wetland communities receive substantial winter grazing by elk and especially bison, a species that may consume large quantities of sedges.

Brichta (1987) studied 21 of the 62 northern-range wetland community types defined by Chadde et al. (1988). He also measured soil types, surface-soil saturation status, and groundwater depth and chemistry. Of the 180 plots studied, 24 mostly were aspen or willow stands in exclosures that excluded herbivores. Depth of soil organic horizon was highly variable, ranging from less than 5 cm for about half of the study plots to more than 60 cm for many sedge-dominated sites.

Community types with abundant sedges had saturated soils all summer. Types with aspen, as well as some graminoid communities, were saturated for part of the summer; community types with spruce, and cinquefoil, as well as other graminoid-dominated communities, were dry most of the summer.

Water tables were variable among study sites. For example, mean depth of water below soil surface was more than 100 cm for spruce wetlands around 100 cm for aspen wetlands, about 20 to 40 cm for willow wetlands and as low as 100 cm, from 40 to 100 cm for cinquefoil wetlands, and for graminoid wetlands from near surface to below 100 cm. Many water tables remained steady throughout the summer; others dropped from near surface to below 100 cm. Several sites with declining water tables were near glacial ponds. Water chemistry was also variable with most pHs near or slightly below neutral. Water chemistry differences did not appear to influence plant diversity, but availability of shallow water through the growing season did. Wetland community types typically changed across moisture gradients.

All of the aspen wetlands and some willow sites studied by Brichta (1987) were within exclosures. At the Junction Butte exclosure, a *Salix geyeriana/ Deschampsia cespitosa* community was inside the exclosure, but a *Potentilla fruticosa/Deschampsia cespitosa* community was immediately outside it. Brichta concluded that "the influences of grazing and succession upon wetland community type distribution should be further studied. Water regimes and soils were similar for paired plots on either side of the exclosures, therefore differences in soils and water levels could not account for the marked difference in vegetation that existed inside and adjacent to exclosures." Brichta's study was during a dry period that eventually resulted in the 1988 fires. Perhaps water tables would have been higher and more sites would have had saturated soils throughout the summer had the study been done during wet years.

Only four species of amphibians are known with certainty to occur in YNP now: boreal toad (*Bufo boreas*), tiger salamander (*Ambystoma tigrinum*), boreal chorus frog (*Pseudacris maculata*), and Columbia spotted frog (*Rana luteiventris*). Two other species have been reported but not verified in recent times. The population sizes of the boreal toad have declined significantly, those of the spotted frog less so, and the chorus frog and tiger salamander appear to vary within normal limits. Wetlands are critical habitat for amphibians, which are often considered indicator species of environmental health (e.g., EPA 1998 Star Grant: Environmental Factors That Influence Amphibian Community Structure and Health as Indicators of Ecosystems). Habitat changes resulting from the interplay between vegetation, hydrology, elk, and beaver could influence available wetland habitat for amphibians, but none of the decreases is clearly related to the known direct or indirect effects of elk population size or feeding.

Conclusions

Hydrological changes in the northern range are the most likely cause of changes in wetland communities, although evidence from exclosures indicates that hydrological factors do not account for the differences between plant communities inside and outside of exclosures. Lowering water tables from stream incision and loss of beaver ponds may reduce wetland habitat. There may be drying of wetland depressions but there are no long-term data on shallow groundwater levels in northern range locations where drying may be occurring to explain changes in these depressions. Some of the depressions may also be filling in, reducing the amount of area available for wetland species (committee observation, Yellowstone National Park northern range, June 1999). Wetlands in the northern range that support herbaceous vegetation may be grazed, but use of these areas probably is not as detrimental to their long-term sustainability as potential changes in groundwater availability. However, wetlands dominated by woody plants appear to be significantly degraded by browsing.

4

Present Conditions: Animals

POPULATION DYNAMICS OF NORTHERN RANGE UNGULATES

YELLOWSTONE'S NORTHERN RANGE supports a rich community of native ungulates. Ungulates in the park were subject to market hunting until the 1880s, and the park's wildlife was not seriously protected until the U.S. Army was assigned administration of the park in 1886 (YNP 1997). By this time, most ungulate populations had been greatly reduced. The Army (and later NPS) managed and protected the resident ungulates and diligently controlled predators, which resulted in greatly reduced populations of coyotes, bears, and mountain lions, and the extirpation of wolves from Yellowstone National Park (YNP). YNP adopted a policy of "natural regulation" in 1968, which led to increased populations of elk and bison. Throughout the twentieth century, management of ungulates has been controversial and great concern has been expressed by the public and park officials about the "correct" management objectives and the actions needed to achieve them.

Density Dependence and Natural Regulation

The concept of density-dependent regulation of population sizes has figured prominently in the controversy over management of elk and bison in YNP

(Houston 1982, YNP 1997). For a population to be regulated by density-dependent factors, some combination of the following processes must operate. As population density increases, mortality and emigration rates increase and the rate of reproduction decreases. Increases in mortality can result from depletion of food supplies because individuals find it increasingly difficult to obtain adequate nutrition. Diseases, whose transmission is facilitated by high population densities, and predators also can increase mortality (Sinclair 1989, Royama 1992, Begon et al. 1996). Rates of reproduction may decrease because females cannot obtain enough food to support high rates of pregnancy and because offspring may be born at lower weights and less appropriate times than when food supplies are good. These rates may change gradually with population density, or there may be thresholds at which major changes occur (Fowler 1987; McCullough 1990, 1992).

The combination of these processes tends to cause population densities to decline when they are high and to increase when they are low. However, this does not guarantee that population densities will stabilize or reach some equilibrium because changes in rainfall, snow accumulation, fires, and other abiotic events may cause large fluctuations in the capacity of the landscape to support the population (Soether et al. 1997). In other words, because the environmental conditions in the landscape may vary considerably, the magnitude of variation in the density of a population by itself cannot be used to assess the importance of density-dependent factors in regulating the size of a population. Most populations of larger herbivores are subject to a combination of stochastic and density-dependent processes that lead to large variation in rates of juvenile survival and subsequent changes in population growth rates (McCullough 1990, Sinclair and Arcese 1995, Soether 1997, Gaillard et al. 1998).

The conceptual basis of density-dependent population regulation is simple, and there are many examples of ungulate populations in which fecundity declines or mortality increases as population density increases. However, no single statistical method identifying density dependence has emerged, despite vigorous discussion (Strong 1986, Pollard et al 1987, Turchin 1990, Dennis and Taper 1994, Soether 1997, White and Bartmann 1997, Shenk et al. 1998, Bjornstad et al. 1999). Many unharvested ungulates are regulated, at some point, by density dependence (McCullough 1979, Sinclair 1979, Fowler 1981, Gaillard et al. 1998), but populations are always subject to a multitude of factors and it can be difficult to distinguish the effects of density from those of other influences. The best evidence for density dependence comes from direct measures of changes in population processes such as mortality, fecundity, and migration (Shenk et al. 1998).

Elk

The intense public debate since the 1920s over management of the northern range elk population has stimulated numerous studies of the Yellowstone elk population (Barmore 1980; Houston 1982; Chase 1986; Merrill and Boyce 1991; Coughenour and Singer 1996a; Singer et al. 1997, 1998). Nevertheless, reliable data on population size and distribution exist for only the past several decades, and population size estimates in published reports differ (e.g., compare Houston [1982] with Lemke et al. [1998]). Elk are highly mobile. In mild winters, they are widely dispersed and make extensive use of forested habitats where they are difficult to count; aerial counts during harsh winters, when elk move to lower elevations, are therefore more reliable than counts during mild winters. Hunting also alters distribution and the number of animals in the population. Aerial surveys of elk initiated in 1956 marked the beginning of reasonably reliable estimates of elk in YNP (Houston 1982). Comparisons of population estimates are confounded by changes in survey technique, differences in the time of survey (e.g., before or after harvest), and vagaries of weather that influence animal movements and visibility.

Harvest and Movement

Historical records of northern range elk illustrate the dominant influence of intense management before 1968 and the effects of management and natural processes since 1968 (Figure 4-1). Public concerns about overgrazing of the northern range resulted in herd reduction by the park at rates that kept the elk population relatively low and stable from the 1920s until 1968 (Houston 1982). After 1967, when elk harvest stopped, the elk population increased (Figure 4-1). From 1968 through 1975, hunter harvests outside the park from the northern herd dropped from about 1,500 elk per year to fewer than 200 (Houston 1982).

The winter late hunt north of YNP resumed in 1967 amid concerns that disturbances due to hunting would inhibit movements of elk from YNP to the historical winter range north of park boundaries. Until the severe winter of 1988, relatively few elk were observed north of Dome Mountain (approximately 16 km north of YNP) (Lemke et al. 1998). The northern range elk population, which had expanded to about 20,000 animals, responded to heavy snows in the winter of 1988-1989 by moving to a lower-elevation winter range en masse. More than 3,000 elk were observed in the area of Dome Mountain,

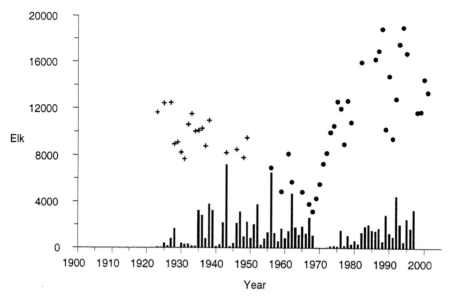

FIGURE 4-1 Elk population counts and harvest of northern range elk. Vertical bars show the number of elk harvested. Filled circles are counts from aerial surveys; crosses are ground counts. Sources: YNP 1997, Lemke et al. 1998, Lemke 1999.

a substantial proportion of the 7,000 to 8,000 elk that migrated north out of YNP that year (Figure 4-2) (Lemke et al. 1998). This event marked a major change, or restoration, in behavior of northern range elk. A significant proportion of the population has consistently migrated to the area of Dome Mountain in subsequent years. From 1975 to 1988, an average of about 200 elk per year wintered north of Dome Mountain; the average increased to about 2,800 per year from 1989 to 2001 (Lemke et al. 1998; Lemke 1999; T. Lemke, Montana Fish, Wildlife and Parks, personal communication, 2001). From 1989 to 1999, an average of 5,600 elk (range, 1,533 to 8,626) wintered outside the northern border of the park, including the area of Dome Mountain (Lemke et al. 1998; Lemke, personal communication, 2001). The many elk wintering outside YNP boundaries are using an expanded winter range. Houston (1982) estimated that during the 1970s the elk winter range consisted of 109,000 ha (as measured by Lemke et al. [1998]), whereas current winter distribution typically includes around 153,000 ha of winter range, an increase of 41% (Lemke et al. 1998). Most of the increase in winter range is outside YNP, where the area utilized increased from 22,000 to 53,000 ha, including 9,200 ha north of Dome

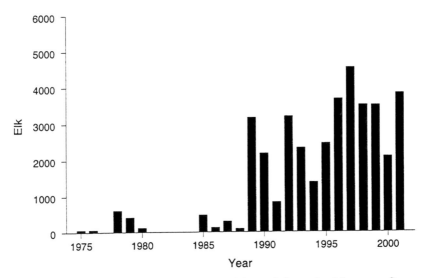

FIGURE 4-2 Number of elk counted north of Dome Mountain, Montana. Sources: Lemke et al. 1998, Lemke 1999.

Mountain. In response to the apparent need for a lower-elevation winter range, Montana Fish, Wildlife and Parks and the Rocky Mountain Elk Foundation collaboratively acquired 3,500 ha of key winter range on Dome Mountain (Lemke et al. 1998). Grazing by domestic livestock was discontinued on this land to provide an enhanced supply of winter forage for wildlife (Lemke et al. 1998).

Regulation of Elk Populations

The northern range elk herd is strongly influenced by both density-dependent and density-independent factors. Mortality of juveniles varies widely from year to year and is positively correlated with population density (Barmore 1980, Houston 1982, Merrill and Boyce 1991, Coughenour and Singer 1996a, Singer et al. 1997, Taper and Gogan 2002). Increased juvenile mortality rates at high elk densities are caused by grizzly bears, black bears, and coyotes (Singer et al. 1997), and, recently, wolves (Smith et al. 1999a). As elk density increases, pregnancy rates (Houston 1982) decline, and a larger proportion of elk calves are born later and at a lower birth weight. These calves survive less well

(Singer et al. 1993), which reduces the rate of population growth (Houston 1982, Merrill and Boyce 1991, Coughenour and Singer 1996a, Singer et al. 1997, NRC 1998). Together, these factors produce a negative correlation between population growth and population size (NRC 1998, Taper and Gogan 2002), a clear indication of density dependence.

Weather has also had a major influence on the current migratory pattern of northern range elk. Deep snow restricts the area available for feeding by ungulates, and their response is to seek better foraging conditions at lower elevations (Coughenour and Singer 1996b). One result of the large migration of elk in response to the severe winter of 1988 was that some elk learned the landscape and continued to leave the park with greater frequency after 1989 (Figure 4-2). Elk migrations out of YNP were clearly influenced by weather. The size of the migration out of YNP is correlated with snow water equivalent (SWE),[1] a rough measure of snow depth (NRC 1998). With more recent data (1989 to 1999), the correlation between the number of elk migrating from the northern range and SWE remains significant ($Y = -2579.1 + 380.8$; $r^2 = 0.41$, $p = 0.03$). Similarly, the number of elk killed in the late Gardiner hunt from 1976 to 1999 is correlated with SWE ($Y = -333.3 + 71.5$; $r^2 = 0.54$, $p = 0.012$). Neither the number of elk leaving YNP nor the take from the Gardiner hunt was correlated with population size ($p > 0.3$). Although regression equations suggest that no elk will leave the park when SWE is less than 5 or 7 inches (12.7 to 17.8 cm), the lowest SWE recorded since 1949 was 10 inches (25.4 cm). Therefore, elk are likely to migrate from YNP in even the mildest of winters.

The postulated effects of severe winter weather on elk have been corroborated by simulation modeling. Landscape-scale simulations of northern range elk identified winter severity as the primary cause of major mortality events (Turner et al. 1994b, Coughenour and Singer 1996b, Wu et al. 1996). Thus, one of the most important factors determining variation in elk population size is one over which managers have no control.

The current pattern of winter use of Dome Mountain may be influenced by artificial feeding. In 1989, Montana Fish, Wildlife and Parks leased agricultural property on Dome Mountain that was used to produce alfalfa. The lease stipulated that only the first cutting of hay would be removed; all subsequent

---

[1]SWE is the sum of snow water equivalent (inches) measured at Lupine Creek, YNP (elevation 2,249 m) and Crevice Mountain, YNP (elevation 2,560 m). SWE data from Farnes (1996) and Farnes et al. (1999).

growth would be left as winter forage for wildlife. From 1992 through 1996, an estimated 209 tons of hay per year was produced on this property (Montana Fish, Wildlife and Parks 1997 lease[2]), a desirable resource within easy reach of northern range elk. Although severe weather may have initially motivated elk movements out of YNP, the existence of high-quality and abundant forage has likely encouraged them to return to the area. A change in the relationship between SWE and elk north of Dome Mountain in winter is apparent from a comparison of data before and after 1989 (Figure 4-3).

A key issue surrounding the natural regulation policy is the size around which the population will fluctuate and the impact of those populations on the northern range. In the absence of wolves, recent estimates of the largest elk populations the winter range can support range from 16,000 to 22,000 animals (Coughenour and Singer 1996a, NRC 1998, Taper and Gogan 2002), a large increase from the early estimates of 5,000 to 11,000 elk (Grimm 1938, 1939; Cooper 1963; Cole 1969). The increases resulted from a better understanding of elk dynamics and from a large increase in the area of available winter range (Lemke et al. 1998). Because only a fraction of the total maximum population is actually observed, these estimates translate to a total population of 20,000 to 22,000 elk that could be supported under the environmental conditions of the past few decades.

Natural Variation in Elk Population Size

Wildlife managers typically prescribe actions that reduce variation in population size or resources. Management of northern range elk has been no different, and observed population fluctuations would likely have been greater in the absence of annual herd reductions, acquisition of winter range, and provision of supplemental forage. Annual harvests prevented elk from attaining high densities that could have exacerbated intraspecific competition, led to more severe nutritional deprivation and adversely affected the range. DelGiudice et al. (1991) found that, during the relatively mild winter of 1987, elk on the northern range showed signs of hunger by midwinter. Nutritional deprivation was significantly associated with declines in cow/calf ratios. During the harsh winter of 1988, winter mortality of elk was severe, and NPS

---

[2]Lease for area designated as Dome Mountain Wildlife Management Area, 1997.

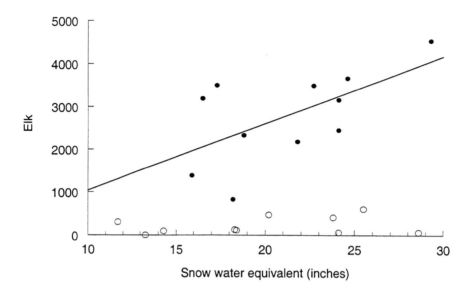

FIGURE 4-3 Number of elk counted north of Dome Mountain, Montana, as a function of SWE. Open circles are data from 1969 to 1988; filled circles are data from 1989 to 1999. Sources: Farnes et al. 1999, Lemke et al. 1998, Lemke 1999.

estimated that more than 4,000 elk died (Singer et al. 1989). Mortality during the 1988-1989 winter may have been increased by the previous summer's drought combined with the large elk population, although a similarly severe winter die-off was reported in 1919 (reviewed by Houston 1982) and less severe die-offs occurred in 1974 and 1996 (YNP 1997; T. Lemke, Montana Fish, Wildlife and Parks, personal communication, January 18, 2000). Large mortality events may be relatively common among mammals, and their periodic occurrence is expected for Yellowstone's ungulates (Young 1994, Erb and Boyce 1999).

Potential Impact of Wolves on Elk

Reintroduction of wolves to YNP in 1995 marked the restoration of the primary predator in the system. Wolves are probably a keystone species (i.e., a species that influences community structure out of proportion to their numbers) (Paine 1966) in the northern range and their activities could touch virtually every aspect of northern range ecology. Wolves regulate herbivore populations in other systems, with consequent effects on landscape and ecosystem processes (McLaren and Peterson 1994, Messier 1994, NRC 1997).

The ability of wolves to regulate Yellowstone's elk population depends on how wolves' consumption of prey changes with prey availability and whether the elk killed by wolves would have died from other causes anyway. Most predators increase their consumption of prey as food becomes more available, thereby reducing the population growth rate of the prey and stabilizing population fluctuations. However, all predators exhibit satiation at some point, and if wolves become satiated when elk are highly abundant, then wolves are likely to have a destabilizing effect, exacerbating population fluctuations caused by severe winters or other factors. An additional, critical consideration is whether wolves kill animals that otherwise would not have died. If wolves kill animals that otherwise would have died from old age or starvation, they add little to the rate of mortality.

Data necessary to develop detailed models of wolf-elk dynamics are unavailable, but simple models have been constructed to evaluate the likely range of effects of wolves on elk, and to a lesser extent, on the northern range ecosystem. Based on early models of wolf-elk dynamics, we could reasonably expect wolves to reduce the elk population by 5-20% (Boyce and Gaillard 1992, Mack and Singer 1993). So far, only Boyce and Anderson's (1999) model includes stochastic variation due to predator behavior and vegetation dynamics. Boyce and Anderson's (1999) model was intentionally simplistic in an effort to make the results interpretable, so their results are most useful for identifying qualitative trends in system responses. Boyce and Anderson compared effects of variation introduced at the bottom (vegetation) or top (predator functional response) of the system. When stochasticity was introduced by varying vegetation production, 95% of the variance in herbivore numbers was explained by vegetation alone. For this model, only 28% of the variance in population dynamics was explained by the number of predators. When variation entered the system at the top, via variation in the functional response of wolves, vegetation and predators alone accounted for 21% and 75% of the variation in herbivore numbers, respectively. These results identify a key problem for future research on interaction of wolves and herbivores: the source of variation in the system can have a profound effect on evaluation of the relative role of regulating factors.

Weather has had a very large impact on dynamics of elk in Yellowstone, and wolves are unlikely to change this. On the other hand, a bad year for elk is likely to be a good year for wolves; therefore, variation in elk-wolf dynamics will almost certainly result from variation due to both changes in forage production and in predation by wolves. Severe weather *per se* is unlikely to have any direct effect on wolves, but severe weather may make it easier for wolves to kill elk, thereby driving greater variation in prey populations (Post et al. 1999).

Bison

Management of bison in and around YNP also has been controversial, but for different reasons. Bison in YNP are infected with the bacterium *Brucella abortus*, the causative agent of brucellosis (Meagher and Meyer 1994), a disease that causes abortion in cattle and is of significant economic and political interest to the livestock industry. Efforts to prevent transmission of brucellosis from free-ranging bison to nearby cattle have resulted in the slaughter of more than 2,000 bison on the boundaries of YNP, which has created a public outcry. The Yellowstone herd is the only free-ranging bison herd that avoided extermination in the late 1800s, and for many people, bison are a symbol of the American West. Americans continue to care passionately about management of Yellowstone's bison. A recent draft management plan (NPS 1998) evoked more than 60,000 written comments from the public.

Bison were widely distributed in North America before YNP was created in 1872, but by 1900, only about two dozen free-roaming bison survived in YNP (Meagher 1973). Bison taxonomy remains controversial, but according to Reynolds et al. (1982) and Meagher (1973) plains bison (*Bison bison bison*) inhabited North America east of the Rocky Mountains; the wood or mountain bison (*Bison bison athabascae*) lived in grasslands in mountain valleys, parks, and northern boreal woodlands and tundra (Reynolds et al. 1982). Yellowstone's remnant bison herd consisted of mountain bison. Park management of wildlife in the early 1900s involved supplemental feeding in winter, protection of bison within enclosures, and culling of weak individuals (Meagher 1973). Bison from domestic herds were transported to YNP to supplement the size of the tiny herd of wild bison, but many of those domestic animals were plains bison. The total number of bison in YNP in 1902 was 44 (Figure 4-4) (YNP 1997). With protection and intensive management, the bison population increased to more than 1,000 animals by the mid-1920s. Harvests were initiated then and conducted most years to keep the bison population at about 1,500 animals until the 1960s. By 1968 the population was 400 (Figure 4-4), after which harvest was stopped (Meagher 1973). In the absence of harvest after 1968, the bison population rapidly and consistently

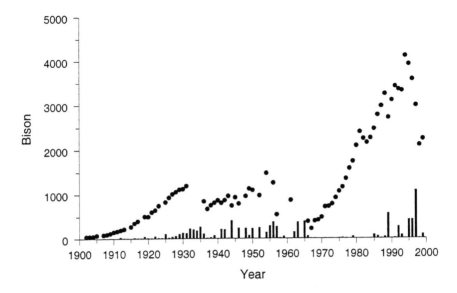

FIGURE 4-4 Total bison count (filled circles) and removals (vertical bars) in Yellowstone National Park. Sources: Dobson and Meagher 1996, YNP 1997, Taper et al. 2000.

increased until the severe winter of 1988, by which time about 3,000 bison lived in YNP. Concerns over transmission of brucellosis from bison to domestic cattle had heightened by 1988, and more than 500 bison were slaughtered as they migrated out of YNP seeking suitable winter range. In the winter of 1996-1997, which had deep snows, more than 1,000 bison were slaughtered when they migrated outside the park.

Bison are gregarious and naturally live in nomadic herds. As the bison population in YNP expanded, bison eventually formed more or less discrete herds. Through 1968, when there were about 400 bison in YNP, bison formed three winter herd subunits and two summer breeding populations (Meagher 1973). Major wintering areas were the northern range, central YNP (Hayden Valley/Mary Mountain), and the Madison/Firehole area in western YNP.

Natural Regulation of YNP Bison

Does the YNP bison population exhibit density dependence, and if so, how many bison is the park likely to support? Relevant data are available for only two periods of protracted growth of the YNP bison population. The first was from 1902 to 1931, when the population grew from fewer than 100 to more than 1,000 animals (Figure 4-4). Analyses of this period are complicated by management actions, which included fencing, artificial feeding, and castration of male calves (Meagher 1973). From 1931 to 1967 the size of the bison population was controlled by intense culling. The second period of protracted growth was from 1968 to 1994, which ended with the killing of more than 2,000 bison between 1994 and 1998. Bison in YNP were never allowed to achieve a population that appears to be in equilibrium, dynamic or otherwise. The analysis of bison population dynamics is further complicated by the nomadic and migratory habits of bison. Given the opportunity, bison travel long distances, and movements of up to 240 km were reported in a northern population (Soper 1941). Hornaday (1889, in Reynolds et al. 1982) documented migratory bison movements of several hundred miles. In YNP, bison clearly responded to increased population density by moving to new areas, both inside and outside park boundaries (Meagher 1989, NRC 1998, Taper et al. 2000). These traits complicate predictions of bison behavior in the Yellowstone ecosystem and suggest that management to restrict bison to park lands will need to be intense.

Given the tendency for bison to expand into new areas as population density increases, identifying density-dependent processes is difficult. Nevertheless, if the bison population exhibited density dependence, its per capita growth rate would diminish as density increases. From 1969 to 1981, the population grew at a more consistent and higher rate than after 1981 (Figure 4-5). Was the diminished growth rate after 1981 solely the result of park herd reductions, or is it an indication of density dependence?

The NRC (1998) examined this question, first focusing on growth of YNP bison from 1972 to 1995, a period with few artificial removals of bison. Over this period, the average annual increment to the population was 145 individuals (NRC 1998), which suggests a continuous decline in per capita reproduction throughout the range of population sizes. This would be a rather unusual pattern of density dependence for an ungulate population (McCullough 1990). The NRC (1998) suggested that this pattern was most likely to arise where female dominance strongly influenced calf survival or where there were few good habitats in which females successfully raised calves.

An alternative interpretation is that since 1968 the bison population has experienced two distinct growth phases. The first phase, from 1968 to 1981, was characterized by rapid and consistent growth rates. The second phase, from 1982 to 1999, has been characterized by high variance in growth rates, density-dependent effects on dispersion and movement, and a much more prominent effect of weather. Most data support this interpretation.

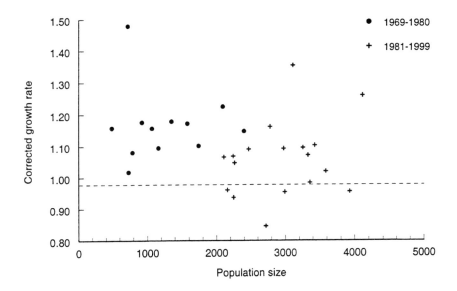

FIGURE 4-5 Growth rate (corrected for removals) of Yellowstone bison. Sources: YNP 1997; Taper et al. 2000; P. Gogan, Montana State University, personal communication, May 12, 1999.

If suitable habitat for bison in Yellowstone was relatively abundant before 1982 and limited thereafter, annual per capita recruitment and growth rates should differ between these periods. Per capita growth rates can be estimated from the annual increment to the population, corrected for harvest. The corrected annual increment to the population is the difference in population size between years, with harvest added back ($N_{t+1} - N_t + H_t$). The average corrected increment does not differ between the two periods (160 vs. 162 bison), but there is a huge difference in variance between the two periods (Figure 4-6). Mean SWE was similar during 1982 to 1999 and 1969 to 1981 (19.9 vs. 22.3 inches [50.55 vs. 56.64 cm]), as was the number of above-average snow years (SWE > 21 inches [53.34 cm]; 8 vs. 7 years). Although this single index of winter severity fails to incorporate many important factors, SWE has thus far been the best predictor of bison movements. It reveals no differences in weather between the two periods, but growth rates were different. Overall, corrected annual growth rate was lower from 1982 to 1999 than from 1968 to 1981 (6% vs. 16%; $p < 0.01$), and the coefficient of variation was greater (0.20 vs. 0.47).

These analyses are subject to the usual uncertainties about population

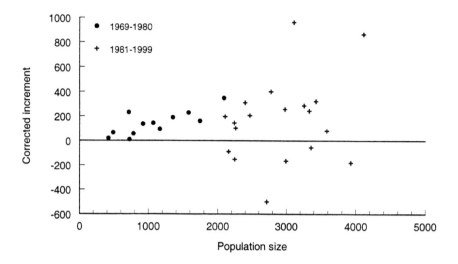

FIGURE 4-6 Annual increment in the YNP bison population corrected for harvest (corrected increment) as a function of population size. Sources: YNP 1997; Taper et al. 2000; P. Gogan, Montana State University, personal communication, May 12, 1999.

estimates, but the trends are too strong to be ignored. They suggest that the Yellowstone bison grows at a constant per capita recruitment until a threshold size is reached, at which point density-dependent factors become important and growth rate, on average, declines (Stubbs 1977, Fowler 1987, McCullough 1990). Such a model suggests that competition has little influence on population processes until resources are largely depleted. For bison this makes sense. Intake rate is an asymptotic function of forage availability (Hudson and Frank 1987, Gross et al. 1993), thus competition is unlikely to influence survival until forage availability is reduced below a threshold level. Reduced survival of calves is most likely to be the first indication of intraspecific competition (Soether 1997, Gaillard et al. 1998), and there is good evidence that this occurs at higher population levels (reviewed by NRC [1998]).

Because weather is highly variable in Yellowstone and influences survival and recruitment, growth of the bison herd is subject to both density-dependent and density-independent influences. At low densities, bison have access to adequate forage in all but the most extreme winters, and their growth rate is stable and high. At high densities, overwinter mortality may be high because good habitat is fully occupied and some proportion of the population is forced into poor habitat, significantly reducing growth rates. As density increases, the importance of both density-dependent (e.g., competition) and density-independent factors (i.e., weather) increases. Both factors contribute to stabilizing the population, but weather tends to add noise to the system and prevent the bison population from stabilizing.

It follows that Yellowstone bison were relatively unconstrained by habitat quality before about 1980. The density of bison in YNP during that period was relatively low, and bison had high and consistent rates of growth (Figure 4-4). Similarly, during the same period the Yellowstone elk population had rapid and consistent growth (Figure 4-1). By about 1980, bison appear to have reached a density where most habitat was fully occupied and, in a typical year, winter habitat was limiting. Bison responded by seeking better habitat, and mortality became density dependent. Data on bison movements are consistent with this hypothesis. Meagher (1989) noted that year-round use of the Blacktail area began in 1980 and that the winter of 1982 marked the beginning of annual use by bison of the area from Mammoth to Gardiner.

Other investigators have also concluded that Yellowstone bison have shown signs of density-dependence limitation. Taper et al. (2000) evaluated the most extensive set of bison observations made by Meagher from spring 1970 to fall 1997. The data came from surveys from more than 160 flights during which the location and size of more than 20,000 bison groups were recorded. Bison used the smallest area during rut and the largest area during winter. As population size and density increased, there was a large increase in the area used by bison during winter, from about 200 km² in 1970 to more than 600 km² in the 1990s. If intraspecific competition were an important factor regulating the size of the bison population, growth rate would be expected to decline as population size increased, after accounting for the effects of area. Taper et al. (2000) thus examined changes in growth rate as a function of density by adjusting for the area actually used by bison in winter. As density increased, growth rate declined ($p < 0.05$), although the predictive value of the regression was very low ($r^2 = 0.29$) and much of the variation in growth rate is not explained by density alone. This suggests that density-independent factors, such as severe weather, were strongly influencing rates of recruitment or mortality.

Taper et al. (2000) estimated the number of bison that YNP might support on a sustained basis. To do so, they determined the area used by bison during winter within YNP (the season when bison are most dispersed) and the density of bison at which the annual growth rate was zero (recruitment = mortality).

These calculations led to an estimated population of about 3,200 bison. Taper et al. (2000) carefully stated that this estimate was a crude approximation and should be used only with great caution. The estimate assumed bison would not leave YNP (i.e., would not migrate as a response to increased density), and it implicitly included the effects of recent management actions that may have modified bison behavior, such as winter grooming of roads and control of bison movements at park borders.

The first response of bison to increased density is to expand the area used, especially during winter. Therefore, one would expect bison to frequently migrate to other areas outside park boundaries. Taper et al. (2000) accounted for this behavior by estimating the density at which bison are unlikely to expand their range beyond park boundaries. That density was approximated by the population of bison in 1984, when management actions began to remove significant numbers of bison at park borders. Bison used about 675 km² of winter range at a population size of about 2,800, which may be the population size at which bison might remain inside YNP without management intervention.

Taper et al. (2000) invoked a rather complicated process to arrive at the estimate of 2,800 bison, but other lines of evidence yield estimates that are nearly the same. The NRC (1998) examined the number of bison removed by harvest at park boundaries as a function of population size. In general, bison begin regularly leaving park boundaries when the population exceeds about 2,500 animals; when the population was more than 3,000, the number of bison leaving YNP was highly related to SWE (NRC 1998). Finally, the bison population numbered about 2,400 animals in 1981, which seems to be a pivotal year in terms of its growth and movement. Taken independently, each of these lines of evidence is weak, but taken together, they provide a consistent picture of the response of bison to increasing density within YNP.

Brucellosis

Because of the importance of brucellosis, much attention and controversy has been directed at the disease (NRC 1998). Brucellosis is endemic in bison and elk herds in YNP. In elk (Thorne et al. 1997) and bison (Williams et al. 1997) brucellosis typically causes abortion of the first pregnancy after infection in most infected females. Epididymitis and orchitis may occur in bison bulls (Williams et al. 1993, Rhyan et al. 1997).

Population consequences of brucellosis on elk and bison on the northern range are not specifically known. Loss of 7% to 12% of the calf crop has been estimated for elk in the Jackson elk herd, where brucellosis is endemic (Herriges et al. 1989, Smith and Robbins 1994). Nevertheless, elk and bison herds infected by *B. abortus* increase at about the same rates as herds without the disease (Peterson et al. 1991, Dobson and Meagher 1996, Williams et al. 1997, NRC 1998), because the effects of brucellosis have been overshadowed by density-dependent factors and climatic effects such as severe winters. Although it is unlikely to be important on the northern range, transmission of brucellosis from elk to bison is suspected to have occurred on the National Elk Refuge (NER) in Jackson Hole, Wyoming (Petersen et al. 1991, Williams et al. 1993, NRC 1998).

Elk on native winter ranges do not appear to maintain brucellosis (Morton et al. 1981, Thorne and Herriges 1992, Toman et al. 1997). Brucellosis was probably present in transplant stock moved from the NER to historical ranges, but with one possible exception (Robbins et al. 1982; E.T. Thorne, personal communication, 2001), it was not maintained.

Population density is important for understanding the dynamics of brucellosis in elk and bison and the likelihood that it will be maintained. At one end of the density spectrum is the situation on the NER and various state feedgrounds where elk are concentrated over artificial feed during the winter and early spring (Boyce 1989, Smith et al. 1997, Toman et al. 1997), when abortions due to brucellosis occur. In severe and even during normal winters, elk are essentially confined to feedgrounds because of limited native winter range or because deep snows preclude them from leaving. Because animals stand body-to-body while eating, a birthing or abortion event of an infected elk could expose large numbers of animals to *B. abortus*. The threshold density of elk required for brucellosis to be maintained within the northern range elk population is not known.

Management of brucellosis will remain a contentious issue, because no management alternative can simultaneously meet YNP's goal of minimizing interference by humans and the livestock industry's goal of brucellosis eradication. Management actions that might eventually eradicate brucellosis are limited to test and slaughter programs and vaccination, but the absence of a highly efficacious vaccine and a way to easily administer a vaccine to free-roaming wildlife are major difficulties. Risk management includes temporal and spatial separation of wild ungulates and livestock and vaccination of nearby cattle (Montana Department of Livestock and Montana Fish, Wildlife and Parks 2000).

Current management of brucellosis in YNP bison relies on the establishment of three management zones and a combination of hazing, test, and slaughter; vaccination when a safe vaccine has been approved for use; and acceptance of some free-ranging seronegative bison outside the park (Montana Department of Livestock and Montana Fish, Wildlife and Parks 2000). The primary purpose of this management plan is to reduce the risk of transmission of brucellosis to nearby cattle by removing the source of infection. The second purpose is to provide negative reinforcement to bison that leave YNP, thereby reducing future risk by diminishing the likelihood that bison will continue to leave YNP. Simulation results consistently show that test and slaughter procedures alone will not reduce the prevalence of brucellosis in bison (Gross et al. 1998).

Other Ungulates

Populations of mule deer, white-tailed deer, pronghorn, bighorn sheep, mountain goats, and moose inhabit the northern range. Because all these species are present in the park in small numbers, the natural regulation policy has not affected them directly. However, the status of pronghorn is sufficiently tenuous that they deserve special attention, regardless of the management of other species.

Mule Deer

Mule deer (*Odocoileus hemionus hemionus*) are the third most abundant ungulate on the northern range. Mule deer populations appear to have fluctuated on the northern range from perhaps fewer than 200 to more than 2,500 individuals in the past 95 years (Barmore 1980). Different counting techniques throughout this period make it difficult to track the population fluctuations with much accuracy. However, it appears that the population both inside and outside the park boundary on the northern range may have grown to 1,000 individuals by 1911 and oscillated near that number until 1987, when the population increased to more than 2,000.

Singer and Renkin (1995) report that minimum mule deer density in lower elevations during the period 1965 to 1968 averaged 4 deer per km². By the late 1980s, minimum deer density averaged 2 per km². Interestingly, Singer and Renkin's data (1995) do not indicate a correlation between minimum mule deer population density and minimum elk population density. Minimum average elk density increased from 6 per km^2 to 16 to 19 per km^2 during the same time period. In contrast, YNP (1997) indicates that mule deer increased during the 1980s simultaneously with increases in the elk population. Singer and Norland (1994) report that mule deer populations increased from approximately 1,000 to 2,200 individuals between 1979 and 1988. Northern range data suggest that the population fluctuated between 1,600 and 2,500 individuals from 1987 to 1999 (Gogan et al. 1999). The population is currently declining and will continue to decline as long as adult female survival is less than 85%, winter survival of fawns is less than 45%, and early winter fawn/100 doe ratios are less than 66. Lemke (1999) reports, however, that segments of the population (e.g., Region 3, east of the Yellowstone River) have increased in the past year.

Approximately 70% of the mule deer in the Gardner Basin migrate out of the park during the winter. The remaining 30% move to winter ranges within the park boundaries at lower elevations (YNP 1997). Some individuals migrate 80 km to their winter range (Wallmo 1978).

Mule deer overlap in habitat use with bison and elk, but they overlap little in their diets (Hudson et al. 1976, McCullough 1980, Singer and Norland 1994). Estimates of percent shrub, determined by microhistological analyses of feces collected on the northern range during the mid-1980s, show that almost 50% of the mule deer diet is composed of shrubs. Elk and bison diets, in contrast, are only 4.1% and 1.6% shrubs, respectively (Singer and Renkin 1995).

Park policy and harvest decisions made in southern Montana are likely to affect mule deer if their population begins to expand. The current levels of harvest and lack of growth suggest that either the harvest is additive (i.e., the animals harvested would not have died otherwise) or that density-dependent factors are becoming strong enough to limit their population growth. Fawn/doe ratios did not decline when populations increased, but ratios have not increased while the population has decreased. This suggests that density-dependent factors are not currently influencing this population. As with other mule deer populations in the Rockies, winter severity, a density-independent factor, probably plays the most important role in determining population fluctuations.

White-Tailed Deer

White-tailed deer have never been abundant in YNP even though they are native to the northern Rocky Mountains. They were apparently abundant in the 1800s in the region around the park, but populations declined until the early 1900s because of overexploitation and habitat change due to forest fires, severe winters (Pengelly 1961, Allen 1971), and heavy browsing of their winter habitat by elk (Murie 1951). Population fluctuations occur, but historical accounts in the Greater Yellowstone ecosystem (GYE) suggest that white-tailed deer have always occurred in low numbers except in the presence of agriculture, and are an "occasional inhabitant" of the northern range (YNP 1997).

White-tailed deer have long been known to exhibit density-dependent population regulation (McCullough 1979). In the YNP region, adult females produce 1.8 to 2.0 fetuses per year, whereas yearling does are less fecund, producing 1.2 to 1.33 fetuses per year. Survivorship of fawns is less than 60% but may be slightly higher in lower elevations and riparian areas. Adult females generally have a survival rate of 65% and a mean life expectancy of 4.2 years. The small population on the northern range suggests that park policy will have little to no effect on white-tailed deer.

Pronghorn

If historical records are accurate, the current population of about 250 pronghorn in the northern range is less than 15% of that in the early 1900s (YNP 1997) (Figure 4-7). Regardless of early population estimates, Yellowstone's pronghorn population is small, isolated, and at severe risk of extirpation (Goodman 1996, YNP 1997, Lemke 1999). Furthermore, northern range pronghorn are of special interest because their population may contain a unique genetic heritage (Lee et al. 1994).

Pronghorn in the northern range occur in lower elevations, primarily along the Yellowstone River from Devil's Slide, north of the park border, to areas east of Tower Junction and up the Lamar Valley. Pronghorn are intolerant of deep snow and their distribution in the northern range is restricted in winter to lower windswept areas where food is exposed throughout the winter, particularly sagebrush grassland (Barmore 1980). The area of pronghorn summer range is considerably greater than suitable winter range, but about 75% of northern range pronghorn are thought to reside in the same area throughout the year, and only about 25% migrate seasonally to a higher-elevation summer range (Caslick 1998).

Estimates of the number of pronghorn in the northern range were up to

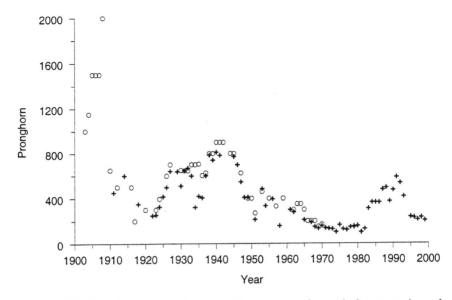

FIGURE 4-7 Pronghorn counts in the northern range. Open circles are estimated population sizes and crosses are actual counts. Sources: YNP 1997, Lemke 1999.

"thousands" during early European settlement, although it appears likely that no more than 1,000 to 1,500 pronghorn actually occupied the area near Gardiner (YNP 1997) (Figure 4-7). The population probably remained at that level until about 1910, after which estimates declined to as few as 200 in 1916 (YNP 1997). By the mid-1920s the pronghorn population recovered and remained relatively stable at 600 to 1,000 animals until 1950. Harvest of 258 animals in 1951 reduced the population to 300 to 400, and it remained at that level until 94 animals were harvested in 1966 (Chase 1986, YNP 1997). Surveys the following year observed only 188 pronghorn (YNP 1997). The harsh winter of 1967 killed about 42% of the herd (Barmore 1980). With the adoption of the policy of natural regulation, harvest of pronghorn within YNP was abolished in 1968. However, the population did not recover from the 1966 harvest and, in fact, declined to 100 to 200 until the early 1980s. Apparently favorable conditions during the 1980s, perhaps related to a series of mild winters, permitted the population to increase to the maximum recent count of nearly 600 in 1991. Over the past decade, the population has again declined, and since 1995 aerial surveys have counted no more than 235 pronghorn (YNP

1997, Lemke 1999). In spite of the small population, permits to hunt northern Yellowstone pronghorn when outside the park have been issued continuously since 1985.

Predation and winter nutrition are the factors most likely to restrict growth of the Yellowstone pronghorn population. Pronghorn in North America generally occupy open prairie grasslands, which may contain a high proportion of sagebrush (*Artemisia* spp.), often a major component of the pronghorn diet (e.g., 78% by rumen analysis, 93% by "utilization") (Bayless 1969). In Yellowstone, the diet of pronghorn on their winter range is similar to the diets of elk, bison, and deer, so these species may come into direct competition during winter (Wambolt 1996). Overlap is less in the other seasons (Schwartz et al. 1977, Schwartz and Ellis 1981). In northeastern California and northwestern Nevada, more than 90% of the winter diets of both deer and pronghorn consisted of woody browse, including sagebrush species (Bayless 1969, Hanley and Hanley 1982).

Pronghorn may be particularly sensitive to severe winter conditions or competition because of their exceptionally high reproductive effort (Byers and Moodie 1990). Pronghorn at the National Bison Range in Montana always produced twins, and the total mass of offspring averaged 17% of the mother's body mass (Byers and Moodie 1990). The rate of twinning by Yellowstone pronghorn is unknown, but pregnancy rates have been high (O'Gara 1968). Nutritional deprivation during winter can cause resorption of one or both fetuses (Barrett 1982) and lead to high overwinter mortality of fawns. Barrett (1982) estimated fawn winter mortality at 53.3% during a severe winter in Alberta. Similarly, severe winters resulted in adult mortality rates of 38%, 50%, and 62% in Saskatchewan, Alberta, and Montana, respectively (Pepper and Quinn 1965, cited in Martinka 1967; Barrett 1982). During a severe winter in Glasgow, Montana, ratios dropped from 90 to 110 per 100 does to 39 to 55 per 100 does at a sagebrush-poor site, whereas at a nearby site with abundant sagebrush there was no such depression (Martinka 1967). Access to prime winter feeding grounds can therefore be particularly critical for pronghorn. In Yellowstone, this was formally recognized in 1932 with the addition of 7,600 acres (3,077 ha) of winter range on the northern edge of the park "as an antelope preserve" (Caslick 1998). The large decline in the pronghorn population before 1961 "would probably have occurred without artificial reductions due to the decline of big sagebrush (*Artemisia tridentata*) in the park and, possibly, cessation of predator control in the park" (Barmore 1980). Big sagebrush is the dominant food item in the winter diet of northern range prong-horn (Singer and Norland 1994).

Pronghorn are the smallest ungulate in YNP and thus they may be especially susceptible to smaller predators, such as coyotes and bobcats, which have difficulty capturing larger ungulates (Byers 1997). Records from the past 40 years indicate that fawn survival rates in Yellowstone are only about 25% (Barmore 1980, Caslick 1998). Caslick attributed most of the mortality to predation, with coyotes, bobcats, and golden eagles accounting for 12% to 90% of fawn mortalities (Caslick 1998). Of the 10 fawns radiocollared in 1991, eight were apparently killed by coyotes within 35 days (Scott 1991). Coyotes are clearly important predators on fawns, and Bruns (1970) specifically noted that coyotes did not appear to prey selectively on wounded or aged animals but fed on all age classes. Byers (1997) documented the extermination of prong-horn on the National Bison Range as a result of predation by coyotes.

Several factors may have increased the density of predators in critical pronghorn breeding areas. By 1998, reintroduced wolves were reported to have killed 6 pronghorn (1 adult and 5 fawns) (Caslick 1998), and wolves may have indirectly increased predation by coyote by displacing coyotes from areas farther from human development to areas where pronghorn live. In addition, late hunts for elk leaving YNP have created a source of abundant food for predators during winter because successful hunters leave animal waste in the field. Predators that consume hunters' waste may switch to pronghorn in the spring.

Bison management may also affect pronghorn because the Stephen's Creek bison facility, where bison are tested for brucellosis and then released or slaughtered, was built in key winter pronghorn range to hold bison that leave the park. Caslick and Caslick (annual reports 1995 to 1999) conducted weekly surveys for pronghorn during winter, recording locations where pronghorn were observed and factors that might influence pronghorn behavior or distribution. They recorded the status of gates (open or closed), observations of predators (coyotes, wolves, and domestic dogs), and disturbances such as operation of the Stephen's Creek bison facility or the presence of visitors in pronghorn habitat. Pronghorn were never present in counting blocks containing the bison facility when it was in operation but were commonly observed in those areas when the facility was unused. Visitors also influenced pronghorn distributions, and pronghorn were much more likely to cross a fenceline via a gate (50% of Scott's 1992 observations) than they were to go under the wire or through buck and pole fences. Scott (1992) also noted that pronghorn were turned away by 28% of fence encounters, whereas pronghorn were detained within the structure of the fence 6% of the time. Because bison movements out of YNP are related to winter snow (NRC 1998), the bison facility is more likely to be used during winters with deep snow, a period when pronghorn are also more likely to be stressed.

Human harvesting of YNP pronghorn has affected this population for more than 50 years, generally correlating with declining population numbers. More than 1,200 pronghorn were removed from the population between 1947 and 1968, a period when the herd declined from 625 to 200 animals. The population continued to decline after hunting ceased and a minimum count of 103 occurred in 1974. By 1985, 365 pronghorn were observed during population surveys, and repeated complaints from local ranchers led to the establishment of a game damage hunt for pronghorn in 1985 (Lemke 1998). The hunt was established by issuing permits to harvest 25 pronghorn annually to control perceived game damage to agricultural crops, but the area was also opened to hunters with a multidistrict bow hunting permit. Although the exact number of pronghorn killed is unknown, estimates are that up to 51 pronghorn per year were harvested from this area (Lemke 1999). In 1995, the number of damage hunt licenses issued each year was reduced to 5, but 12 animals were taken the following year (bow hunters took the balance). The population has again declined to about 200 animals. Effective in fall 2000, Montana Fish, Wildlife and Parks restricted hunting of this pronghorn population to five annual damage permits for an early season. Because few pronghorn use this area during hunting season, the projected harvest is one to three animals per year (T. Lemke, Montana Fish, Wildlife and Parks, personal communication, June 14, 2001).

The ability of the herd to recover from removals and hunting depends largely on factors that affect the production and survivorship of fawns. Fawn production depends on the ability of female pronghorn to maintain good condition, particularly during the winter, and this depends on the severity of the winter and the availability of food. An increasing elk population is reducing the availability of sagebrush, a key winter forage for pronghorn. In addition, winter use of YNP by people has dramatically increased, first because of snowmobilers and more recently because of wildlife watching, especially for wolves. This pronghorn herd faces a serious risk of extinction. The risk could be reduced by eliminating harvest, by increasing harvest of coyotes and bobcats, by restricting recreational access to critical winter range, by reducing disturbances within the winter range, and by management actions that result in enhanced growth and vigor of sagebrush within winter range. Survival of fawns begins with the condition of their dams but is most critically determined by predation pressure. Predators currently appear to be major contributors to fawn mortality.

Moose

Moose inhabit riparian and forested areas from northern Colorado through Alaska, where they typically feed on shrubby plants, especially willows. Many of the preferred foods of moose are early-successional plants; thus, disturbances such as fire and floods may be important for maintaining a suitable habitat. In general, moose are not migratory, although summer and winter ranges may differ in elevation. Moose are reputed to be the only ungulate in YNP whose winter range is higher than its summer range. Moose tend to spend most of the year as solitary animals or as female-offspring pairs (Miquelle et al. 1992).

Although relatively little is known about the ecology of moose that inhabit the northern range, the population clearly is small. Recent surveys resulted in estimates of about 200 animals, but the accuracy of the estimates is uncertain (YNP 1997). There are currently no surveys for moose because of the difficulty and expense associated with attempting to estimate the size of a widely dispersed population that inhabits heavily forested and rugged terrain.

Yellowstone's northern range probably contains relatively little suitable habitat for moose. Tyers (1981) found that about half the forage consumed by northern range moose consisted of subalpine fir and lodgepole pine, their preferred winter forage. He found that most browsing took place in old forests. Although moose are large, they are selective feeders that choose higher-quality forage during summer and they prefer deciduous to conifer browse (Belovsky 1981, Renecker and Hudson 1988, Shipley et al. 1998). In the absence of predators, moose prefer summer riparian habitats with abundant willow. These observations suggest that moose in the northern range are surviving in marginal habitat and that the northern range is unlikely to support a substantial moose population. The abundance of moose has probably declined with the loss of willows.

Reintroduction of wolves is likely to affect moose, but because so little is known about northern range moose, we are unlikely to be able to tell whether wolves have any effect on them. In other areas, moose populations have been regulated by the combined effects of predation by wolves and bears (reviewed by NRC 1997). Elk are likely to remain the primary prey of wolves, but the existence of a high-density wolf population presents a high risk to moose, especially calves.

Mountain Goats

Mountain goats were introduced by Montana Fish, Wildlife and Parks to the Absaroka Mountains northeast of YNP; they are exotic to YNP (Laundre 1990). As the introduced population increased, it expanded and colonized habitats on the border of YNP and, to a limited extent, within the park. Although diets of mountain goat and bighorn sheep overlap substantially (Laundre 1994), these species tend to live in very different habitats in YNP (Varley 1994, 1996). Outside park boundaries, mountain goat populations are controlled by hunting. The area of suitable habitat within the park is limited (Laundre 1990). Mountain goats are a management concern because in some habitats, especially in the absence of large predators, they have the potential to increase very rapidly (Hayden 1989) and achieve densities that may result in habitat degradation (Pfitsch et al. 1983), but this does not yet appear to have occurred in YNP.

Bighorn Sheep

Historical accounts suggest that the bighorn sheep population on the northern range was formerly much larger than in recent times (Schullery and Whittlesey 1992). The causes of the decline before 1981 are unknown but may include disease transmission from domestic livestock or competition with other ungulates. Although there is substantial diet overlap between elk and bighorn sheep (Houston 1982, Singer and Norland 1994), the species tend to use different habitats, and the extent of competition that might occur is unknown.

Small bighorn sheep populations can be highly susceptible to a variety of diseases that can result in catastrophic population crashes (Bunch et al. 1999); an epidemic of keratoconjunctivitis (pinkeye) occurred in the northern range bighorn population in 1981. This event was associated with a major population decline: bighorn counts declined from about 500 to fewer than 200 animals (Figure 4-8). Small bighorn sheep populations may be highly susceptible to

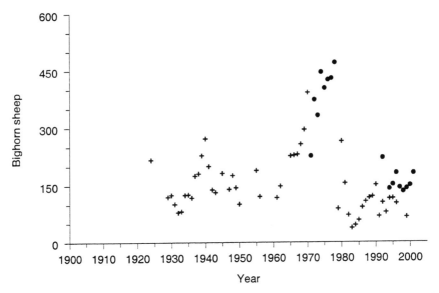

FIGURE 4-8 Bighorn sheep counts in the northern range. Filled circles are aerial counts and crosses are ground counts. Note that estimates from the methods differ, and the apparent large changes in population size are partly due to different survey methods and area surveyed. Sources: YNP (1997) for details of surveys; data from YNP (1997) and Lemke (1999).

regulation by predation (Wehausen 1996, Ross et al. 1997), and ongoing studies of mountain lions may provide a better indication of the relative importance of factors that may control the bighorn sheep population.

MAJOR PREDATORS IN THE YELLOWSTONE ECOSYSTEM

Four large terrestrial carnivores prey on Yellowstone's ungulates: grizzly and black bears (*Ursus arctos* and *U. americanus*), mountain lions (*Puma concolor*), and wolves (*Canis lupus*). Coyotes (*Canis latrans*), medium-sized carnivores (Buskirk 1999), also prey on ungulates. Humans also kill substantial numbers of ungulates and may be considered a sixth major predator in the GYE.

Large terrestrial carnivores have large home ranges and move long distances, often ignoring political boundaries. For example, the range of Yellowstone's grizzly bears "links most of the habitats, and associated species of the GYE" (Clark et al. 1999), and mountain lions radiocollared in the northern

GYE have dispersed to distant areas in Idaho, Wyoming, and Montana (Murphy 1998). As with most other mammals (Greenwood 1980), young males of these terrestrial carnivores typically disperse farther than females (Craighead et al. 1999).

Large carnivores typically utilize a variety of habitat types, and although they may specialize on particular species or sizes of prey, they consume a variety of prey and other foods as the opportunity or need arises (Johnson and Crabtree 1999). Bears eat primarily plant material, although Yellowstone's grizzlies consume relatively more ungulates than most other grizzly populations (Jonkel 1987, Mattson et al. 1991, Knight et al. 1999). Large carnivores and their prey are intelligent and adaptable animals (Berger et al. 2001), which makes their interactions complex.

Among the influences on populations of large carnivores are changes in the abundance and relative abundance of different prey species; natural forces such as fire and weather, which influence the abundance, distribution, and availability of prey; human activities such as habitat alteration and hunting of prey populations; and changes in the abundance and behavior of other prey and predators.

Large carnivores inhabited Yellowstone long before the first European explorers arrived. Bones of wolves, coyotes, and grizzly bears were found in various strata during excavations at Lamar Cave, whose strata extend back about 3,000 years. Multiple Euro-American observers reported wolves, coyotes, grizzly and black bears, and mountain lions in the GYE before 1882 (Schullery and Whittlesey 1999).

Historical Human Impacts on Carnivore Populations

Predator eradication was a major goal in late nineteenth- and early twentieth-century America. Predators were easily poisoned with strychnine-laced carcasses, and by 1880, wolf, coyote, and mountain lion populations in the GYE were greatly reduced (Schullery and Whittlesey 1999). When the U.S. Cavalry arrived in 1886, predators were at first protected along with other animals. However, to protect game species, poisoning of coyotes was resumed in 1898, and in 1907 army personnel were directed to kill coyotes, wolves, and mountain lions (Schullery and Whittlesey 1999). From 1904 to 1935, predator control in Yellowstone resulted in the killing of at least 4,352 coyotes, 136 wolves, and 121 mountain lions (Schullery and Whittlesey 1999). However, by the 1930s, there was mounting opposition among ecologists and conservationists to predator control in national parks. In 1936, official park policy changed to protect native predators, including coyotes, although park managers could still kill individual predators deemed "harmful" (Schullery and Whittlesey 1999). However, wolves were effectively extinct in Yellowstone by then (Smith et al. 1999b) and predator control probably had eradicated mountain lions as well (Craighead et al. 1999).

Bears fared better than wolves and mountain lions. Bear populations were probably reduced by widespread poisoning of ungulate carcasses and by reductions in ungulate populations due to uncontrolled market hunting in the 1870s (Schullery and Whittlesey 1992). However, once they received army protection in 1886, bears began to feed at garbage dumps near hotels, where they soon became a popular tourist attraction (Schullery 1992, Knight et al. 1999).

YNP personnel supported and expanded the bear feeding program and bears ultimately were fed at numerous garbage dumps as well as along roadsides. Grizzlies dominated the dumps, although male black bears used them when grizzlies were not present; female and subadult black bears tended to beg for food from tourists on park roads (Knight et al. 1999). Large numbers of bears became habituated to humans, and they injured people and damaged property. These "problem" bears were then removed from the park. From 1930 to 1969, 46 people were injured by black bears, and an average of 24 black bears were removed from the park per year (Schullery 1992). Although the viewing of bears feeding at dumps was immensely popular with tourists, opposition to the practice grew and the park responded by closing the last public-viewing area at a dump during World War II, although it did not actually close the dumps until the late 1960s and early 1970s (Knight et al. 1999). Closing the dumps led to very high grizzly bear mortality; 229 grizzly bears were removed from the GYE between 1967 and 1972 as bears that previously fed in dumps began to seek food in campgrounds and threatened human safety (Knight et al. 1999). Black bears begging for food remained a common sight along roads until the late 1960s. By 1975, park managers had effectively eliminated this sight by improving sanitation, enforcing a no-feeding policy, and removing begging bears (Knight et al. 1999).

The states surrounding Yellowstone gradually reduced or prohibited hunting of grizzly bears, beginning with a prohibition by Idaho in 1946. However, a complete moratorium on hunting grizzly bears anywhere in the GYE was not imposed until 1974. The GYE grizzly bear was listed as threatened under the Endangered Species Act in 1975 (Knight et al. 1999).

Current Population Dynamics of Large Carnivores

Grizzly Bears

Yellowstone's grizzly bears are the best-studied bears in the world, yet their status and future prospects continue to be subjects of vigorous controversy. Obtaining accurate estimates of the size and dynamics of a grizzly bear population is inherently difficult. Grizzlies are long-lived and females do not reproduce until they are about 6 years old. Grizzly bears are not easily observed as they are mostly solitary and travel over large home ranges in remote, mountainous, and largely forested country. There are additional logistical difficulties in reaching many parts of the study area, recapturing animals, and maintaining operative radiocollars on individual females to obtain long-term reproductive data.

Grizzly bear research began in YNP in 1959. An interagency Grizzly Bear Team was established in the early 1970s. At first, park authorities added to natural logistical difficulties by prohibiting radiocollaring of bears until 1975; then they allowed only one or two bears per year to be collared (Knight et al. 1999). By 1982 there were enough data for researchers to conclude that adult female mortality was high and that the reproductive rates were so low that the population was declining at about 2% per year (Knight and Eberhardt 1985). In 1983, an Interagency Grizzly Bear Committee was formed and charged with devising management strategies to reverse the population decline. Agencies began to manage habitat for grizzlies by eliminating sheep allotments within grizzly areas, increasing efforts to prevent illegal killing of bears, and changing policies such as food storage, garbage disposal, and removal of problem bears to minimize the need to legally kill bears (Knight et al. 1999).

Population analyses suggest that the Yellowstone grizzly bear population was relatively stable from about 1959 to 1993, with periods of slight increase or decrease (Eberhardt et al. 1994, Pease and Mattson 1999, Boyce et al. 2001). Pease and Mattson (1999) predicted a large (~15%) decline in the size of the grizzly bear population from 1993 to 1996 because of widespread failure of whitebark pine, a key food resource. Other models were less sensitive to this factor and suggested a slight increase in the size of the grizzly bear population over the same period. Counts of individual females with young cubs and survival rates also indicated a positive trend in grizzly bear numbers (Knight et al. 1999).

All recent estimates of the size of the Yellowstone grizzly bear population are in the low hundreds (Craighead et al. 1999). These include an estimate of

390 based on marked females; an estimate of 339 based on known families of bears; a statistical (bootstrapped) estimate of 344 with a 90% confidence interval of 280 to 610 bears (Eberhardt 1995); and an interagency review committee estimate of at least 245 bears, of which 67 were adult females (Eberhardt and Knight 1996).

The Yellowstone grizzly bear population is currently isolated from other grizzly populations and is not large enough to avoid loss of genetic variation in the short term (Harris and Allendorf 1989). The current genetic effective population size is only 13 to 65 bears (Paetkau et al. 1998). The genetic effective population size of a wild population, which is generally much less than the census size of the wild population, is defined as the size of a hypothetical population that would have the same rate of decrease in genetic diversity by genetic drift (or increase in inbreeding) as the focal wild population (Hedrick 1983). Loss of genetic diversity reduces population fitness and the probability of long-term survival; thus YNP's grizzly bears probably need more protected habitat and dispersal corridors to preserve genetic diversity (Craighead et al. 1999). The best habitat for possible future population expansion appears to be in Wyoming. The greatest single threat to the population is increasing development of private lands, which not only decreases habitat but also greatly increases the potential for human-bear conflicts and consequent death of the bears involved (Knight et al. 1999).

Grizzly bears require a diverse habitat with minimal human disturbance to cope with climatic changes, alterations in the availability of different foods, human impacts, and changes in the abundance of other wildlife populations. Grizzlies can exploit marginal habitat to some degree but they require time to learn new habitat-use patterns when conditions change (Jonkel 1987).

In Yellowstone, grizzlies feed on weakened and winter-killed elk and bison from March through May, and they kill newborn elk calves during May and June (Singer et al. 1997, Knight et al. 1999). A few individual bears kill healthy elk during the summer, and bull elk become more susceptible during the fall rut. Seeds of whitebark pine (*Pinus albicaulis*), currently threatened by white-pine blister rust (*Cronartium rubicola*) in the GYE, are an important food for Yellowstone's grizzly bears. About 30% of the most productive whitebark pine areas burned during the 1988 fires (Knight et al. 1999).

Black Bears

Very little is known about the numbers and population dynamics of Yellowstone's black bears; research has been directed disproportionately at grizzly bears. The population has probably decreased since dumps were closed and feeding bears along roadsides was stopped. However, the only study of Yellowstone's black bears was conducted more than 30 years ago. Barnes and Bray (1967) estimated a minimum density of 0.07 bear per km^2 in their Gallatin Mountain study area. Estimates of black bear densities in other areas of North America range from 0.1 to 1.3 bears per km^2; however, densities are generally lower at higher altitudes (i.e., YNP) because of the shorter foraging season and poorer soils (Kolenosky and Stratheam 1987). Cole (1976) estimated that there were 650 black bears within YNP by extrapolating the Barnes and Bray estimate over a larger area. Craighead et al. (1999) estimated that there are currently fewer than 2,000 black bears in the GYE. The status of the population is uncertain because of the lack of data, although the population is generally presumed to be stable. About 1,000 black bears are legally killed in Montana each year (Craighead et al. 1999).

The ecology of black bears is known from many studies in other areas (Kolenosky and Stratheam 1987). Black bears are habitat and feeding generalists (Johnson and Crabtree 1999). They are usually forest dwellers, and the best black bear habitat is mixed forests that contain a variety of tree and shrub species of different ages (Kolenosky and Stratheam 1987). Black bears in Yellowstone typically prefer spruce-fir habitats and adjacent meadows but were often observed in lodgepole-pine forests along roadsides during the era of roadside feeding (Barnes and Bray 1967). Black bears are basically vegetarians and their diet appears to be largely determined by local food availability. Bears adapt to new sources of food and change their foraging habits accordingly. Black bears can kill young ungulates, which are vulnerable for a 2 to 4-week period after birth (Kolenosky and Stratheam 1987), and also eat carrion and insects. Killing young ungulates appears to be a learned behavior, and once learned, may continue to be part of an individual's foraging routine (Kolenosky and Stratheam 1987). Carrion provided by cougars and other predators may be a significant food source for black bears in Yellowstone (Crabtree and Sheldon 1999a, Murphy et al. 1999).

Ecological conflicts exist between black and grizzly bears and may be depressing populations of both species (Jonkel 1987). However, interactions between Yellowstone's black and grizzly bear populations have not been studied. Where both species coexist in Alaska, grizzly bears dominate black bears (Miller et al. 1997). In interior Alaska, grizzlies are most commonly associated with alpine tundra habitats, whereas black bears frequent forest and lowland areas (Klein et al. 1998). In the absence of grizzly bears in northern Quebec, black bears have moved into tundra habitats normally occupied by barren ground grizzlies west of Hudson Bay (J. Huot, Faculty of Science and Genetics, University of Laval, personal communication, January 18, 2001).

Mountain Lions

Mountain lions have excellent long-distance dispersal abilities (Murphy et al. 1999). Young males have dispersed as far as 480 km from their natal home range (Craighead et al. 1999). Presumably, these dispersal abilities enabled mountain lions to recolonize YNP after they were locally eradicated by predator control programs. The GYE mountain lion population ranges over some 2,200 ha of relatively contiguous habitat (Murphy et al. 1999). Craighead et al. (1999) estimated that there are fewer than 500 adult mountain lions in the GYE, but they gave no basis for this estimate. Because of their high female survivorship and fecundity (litters of up to six kittens), mountain lion populations are more resilient than those of some other large predators, such as bears (Weaver et al. 1996).

Human-caused mortality is an important influence on most mountain lion populations, including those in the GYE outside the boundaries of YNP (Murphy et al. 1999). Most of this mortality is due to legal hunting: 48% of mortality among adult and subadult radiocollared mountain lions in the northern Yellowstone ecosystem was due to hunting; another 48% was attributed to natural causes (Murphy et al. 1999). Approximately 500 mountain lions were killed by hunters in Montana in 1998 (Craighead et al. 1999). However, mountain lion populations appear to have stabilized or even increased in many areas of the northern Rockies, including in the northern range area (Craighead et al. 1999), despite hunting pressure (Murphy et al. 1999).

Mountain lions occupy a wide range of habitats, although they prefer areas of steep and rugged topography (Lindzey 1987). Abundant cover is important to them as it provides security from enemies, including other predators and humans, and increases hunting success as mountain lions typically hide and ambush their prey. Thus, the structural characteristics of the local vegetation appear to be more important than the dominant plant species (Lindzey 1987).

Mountain lions are almost totally carnivorous and can kill all the YNP ungulates except adult bison. The diet of mountain lions varies with the abundance and availability of prey seasonally and geographically. Deer compose a major portion of their diet in most areas, although mountain lions also kill large numbers of small prey, such as snowshoe hares (*Lepus americanus*) when they are abundant (Lindzey 1987). In the northern GYE, elk calves are a major source of food for mountain lions (Murphy 1998). However, because there are relatively few mountain lions and they kill far fewer ungulates than human hunters, Murphy (1998) concluded that they have little direct effect on the size of the elk and deer populations. Mountain lions are responsible for about 3% of the elk and 4% of the mule deer deaths in the northern GYE (Murphy 1998). Mountain lions kill about 12% of the buck mule deer, 9% of the elk calves, and 1% of the bull elk, but less than 5% of the other age-sex classes of elk and deer.

Wolves

Wolves in the northern Rocky Mountains were listed as endangered under the federal Endangered Species Act in 1974. Many scientists favored reintroduction over natural recolonization as a means of restoring wolves to the GYE. However, because of extensive controversy, wolves were not reintroduced to YNP until more than 20 years after they were listed as endangered (Bangs et al. 1998). In 1995, 14 wolves in three packs captured in Alberta, Canada, were introduced to YNP. In 1996, another 17 wolves in four packs captured in British Columbia, Canada, were introduced to YNP. In 1997, 10 pups and 3 adults from a pack captured in northwestern Montana (because they were chasing livestock) were released in the park (Bangs et al. 1998).

The wolves have thrived and are being extensively monitored. Ten packs in 1997 and seven packs in 1998 produced pups; by fall 1998, the population estimate was 116 wolves (Bangs et al. 1998). According to the Yellowstone wolf project report for 1998 (Smith 1998), litter size in 1998 averaged 5.5 pups and 81% of the 44 pups born survived to the end of 1998. Fifteen wolves died in 1998: five pups, four yearlings, and six adults. About half this mortality was due to natural causes, including wolves killed by other wolves, avalanches, and elk. The remaining mortality was due to human activities, including wolves killed by control actions, illegal shooting, and vehicles. Some assumptions made before the initiation of the wolf reintroduction program have been validated (Smith et al. 1999b): no preexisting wolves were found in the GYE and it was necessary to impose land-use restrictions around wolf dens. However, wolves have killed more than the predicted 120 ungulates per year and fewer than the predicted 19 cattle and 15 sheep per year. Many more visitors than expected have seen wolves.

It was predicted that wolves would travel outside the experimental area of

YNP and immediately adjacent neighboring GYE lands, but this had not occurred through the end of 1998 (Smith et al. 1999b). Wolves regularly disperse long distances: male wolves disperse a mean of 85 km and a maximum of 917 km from their natal home ranges (Craighead et al. 1999). Some wolves had dispersed out of the park as far south as the Jackson area (which is still part of the primary wolf recovery area) (Smith 1998), but none had dispersed from the area designated as the "Yellowstone Nonessential Experimental Population Area," which includes all of Wyoming, Montana as far north as the Missouri River, and part of Idaho (Bangs et al. 1998).

Wolves in different geographical locations rely on different species of prey. Wolves in North America prey largely on ungulates and beavers, but also take other types of prey (Carbyn 1987). Elk have been the wolves' major prey in the GYE to date, which is not surprising given the abundance of elk. However, the wolves are known to have killed every ungulate species present except for bighorn sheep (Smith et al. 1999b). In 1998, wolf researchers found 109 definite and 121 probable wolf kills (Smith 1998). Eighty-six percent of these kills were elk (198), followed by 3% each of mule deer (7), pronghorn (6), and coyotes (7); 2% of bison (5); and 1% each of moose (3), wolves, and unidentified prey. The age composition of the elk kill was 43% calves, 21% cows, 19% bulls, and 16% unknown. Packs on the northern winter range killed an average of one ungulate every two to three days during March and one every three to four days during November and December.

Wolves can adjust to changing prey abundance and vulnerability (Messier 1995), although in multiple prey systems they may be slow to change their favored prey species (Dale et al. 1995). In some instances, wolves have switched to different age classes as their proportion in the prey population changed as a consequence of previous predation (Dekker et al. 1995) or winters with deep snow (Mech et al. 1995). Wolves have also learned to kill new types of prey (Klein 1995). However, it is difficult to predict how or when there may be substantial changes in the prey taken by Yellowstone's wolves.

The Yellowstone wolves have been returned to an ecosystem that existed without them for much of the past century. Adjustment of the system to the presence of wolves is likely to take many years, as the wolves and the other components of the ecosystem adjust to one another (Klein 1995, Berger et al. 2001). Wolves will probably alter their patterns of prey selection, pack structure, movements, and population dynamics as the density, distribution, predation avoidance behavior, and population structure of their prey species change. A similar adjustment can be expected between wolves and other carnivores. A return to relative stability of predator-prey relationships within the Yellowstone ecosystem may take longer than the period that wolves were missing from the system. Long-term monitoring of the predators and ungulates within the Yellowstone ecosystem provides a unique opportunity to greatly expand knowledge about interspecies relationships among upper trophic levels.

Coyotes

Coyotes are the smallest and most numerous of Yellowstone's major predators. Although they were the first of the park's predators to be intensively studied (Schullery and Whittlesey 1999), there was a long gap between the publication of Adolph Murie's classic monograph on coyotes (1940) and the modern era of coyote research. The latter began in 1989, six years before wolves were reintroduced to the ecosystem (Crabtree and Sheldon 1999a). Reintroduction of wolves was expected to have a negative effect on Yellowstone's coyote populations, as larger species of canids tend to compete for prey with, kill, and physically displace their smaller competitors.

Crabtree and Sheldon (1999a, 1999b) recently summarized data collected during these new studies of the coyote on Yellowstone's northern range. Coyotes live in all GYE habitats below about 2,400 m except for very steep, rocky areas and areas with deep snow. Crabtree and Sheldon (1999b) estimated an average pre-wolf density of 0.45 adult coyote per km^2 on the northern range and densities from 0.1 to 0.4 coyote per km^2 over much of the forest habitat in the GYE. Higher densities, sometimes exceeding 1 coyote per km^2, occur in some of the more open grassland and shrub habitats. Craighead et al. (1999) estimated that there are fewer than 3,000 adult coyotes in the GYE.

Yellowstone's coyotes are protected from hunting and trapping and inhabit an environment with good food resources. Crabtree and Sheldon (1999a, 1999b) argued that these factors explain many of the differences between Yellowstone coyotes and those studied in other areas, including larger pack size, greater social stability, higher adult survival rates (91% per year), higher mean age of adults, and lower dispersal rates of juveniles.

Coyotes on the northern range have a social system similar to that of wolves, living in territorial packs with an average size of six adults. However, unlike wolves, members of coyote packs often travel alone. Coyote territories are contiguous (no intervening spaces between them), nonoverlapping, and average 10.1 km^2. Territory boundaries in the Lamar Valley and the Blacktail Plateau were very stable from 1990 through 1995, changing little from year to year.

Even though female coyotes become sexually mature around 10 months of age, Yellowstone females do not breed until they are 2 to 5 years old. An average of 5.4 pups are born per territory but an average of only 1.5 pups survive to 1 year of age. Principal causes of pup mortality are disease and starvation.

Coyotes feed primarily on voles and elk carcasses and the diet varies seasonally (Murie 1940, Crabtree and Sheldon 1999b). Ungulates provide about 45% of the coyotes' annual biomass consumption, most of which is consumed as carrion during the five winter months.

The reintroduction of wolves in 1995 had dramatic effects on the coyote population (Crabtree and Sheldon 1999b). Wolves killed from 25% to 33% of the coyote population during each of the 1996-1997 and 1997-1998 winters, especially in the core areas used by wolves. Almost all these coyotes were killed near the carcasses of elk killed by wolves. The coyote population in the Lamar Valley dropped from 80 coyotes in 12 packs in 1995 to 36 coyotes in 9 packs in 1998 (Crabtree and Sheldon 1999b), and coyotes failed to recolonize their traditional territories in core wolf areas.

However, coyotes are flexible and adaptable animals and have already begun to travel in larger groups (Crabtree and Sheldon 1999b). Some of the surviving coyote packs are smaller and are producing larger, healthier pups with higher survival rates. Coyote packs on the fringes of wolf territories have experienced little mortality and are able to benefit from elk carcasses killed by neighboring wolves.

5

Conclusions and Recommendations

NATURE IS DYNAMIC

A PERVASIVE THEME in this report is the dynamic nature of the northern Yellowstone ecosystem. Over long periods, a changing climate and major geological processes have resulted in dramatic restructuring of the landscape and associated plant and animal communities. The Greater Yellowstone Ecosystem (GYE) has experienced large-scale disturbances including fire, floods, blow downs, ungulate and predator population fluctuations, and outbreaks of diseases and insects that affect plants and animals. In addition, during the late 1800s, intense reduction of carnivores and ungulates diminished or eliminated populations of key species. Furthermore, the northern range is part of a larger system where human activities are steadily increasing. Thus, we probably cannot ever manage Yellowstone National Park (YNP) to maintain some agreed-upon stable condition, if that were to become a management objective. We lack sufficient knowledge, resources, and capability to sustain any environmental state through active management.

Given the ever-changing nature of the northern range on both temporal and spatial scales, can we determine which of the changes we observe in ungulate numbers and range, forest conditions, and riparian conditions are within the bounds of natural variation and which, if any, are caused by human activities?

To answer the question we must assess ecosystem resilience, resistance, and stability. Is the system easily modified? Does it readily recover from perturbation? Are there thresholds that result in major or irreversible changes in processes, ecosystem conditions, or population numbers?

Theory and field studies have shown that some ecological systems change abruptly from one relatively stable state to another. In these situations, simply removing the factor or factors that caused change may not return the system to its previous state. For example, sustained, heavy livestock grazing in arid grasslands of the western United States, in the absence of fire, has led to invasion and establishment of shrubs and trees (Archer 1994). Once trees gained sufficient stature to capture much of the moisture supply, elimination of grazing did not result in reestablishment of grassland (Glendening 1952). Such a process is consistent with "state and transition" models and with the existence of multiple stable states (Allen-Diaz and Bartolome 1998). These conceptual models help us to appreciate the complexity of ecosystem relationships and processes and should be used to evaluate management of the northern range. How do these concepts help us to evaluate changes in the GYE? Many aspects of the northern range have been intensively studied, but it has not been experimentally shown, for example, how large a reduction in the consumption of aspen by ungulates would be required to permit their "recovery." Consequently we do not know whether changes in plant communities during the 1900s indicate that a new state, characterized by fewer communities dominated by willows and aspen, is likely to persist. Research outside the park, however, does not support the hypothesis that a new state has become established (Kay 1990). To evaluate whether the northern range is approaching a threshold, beyond which willow and aspen communities will be unable to reestablish themselves, we must have some idea of the range of natural variation (Landres et al. 1999). Are changes on the northern range within limits to be expected since Europeans arrived? How important are rare events? The "natural" interval between large fires is thought to be on the order of 200 to 300 years—can we realistically expect to manage such events? Despite claims to the contrary, we found no evidence that the northern range is approaching a threshold after which we would observe irreversible changes, such as loss of local reproductive potential of key plant species (e.g., sagebrush or aspen), that would not have occurred if the park actively controlled ungulates. This finding results, in part, because much of the evidence of dramatic changes comes from communities that are successional or the result of disturbance (e.g., aspen and riparian communities) (Houston 1982). However, changes in sagebrush cover and grassland composition, vegetation types that are neither successional nor the result of disturbance, have also occurred.

In view of the profound changes that have occurred around the GYE, it is no longer possible to have an ecosystem that is identical to the natural state that existed there before European settlement—that is, containing about the same numbers and distributions of all the species of plants and animals. YNP still has all the species present there 150 years ago, but many of the large mammals can no longer respond to change as they used to—through migration or dispersal (Wambolt and Sherwood 1999). No aspect of the ecosystem can be considered "natural" in that sense. The question is whether the ecosystem appears to be headed for some state that is very different from any previous state that we know about in the past few thousand years. We do not think it is. Vegetation changes observed in the past 130 years or so appear to have been influenced more by ungulate browsing than by climate change.

MANAGEMENT FOR ECOSYSTEM STATE OR ECOSYSTEM PROCESS?

Natural resource managers typically try to reduce variation around some desirable ecosystem state. For wildlife managers, a desirable state usually is defined by a consistent harvest of the target species, stable vegetation communities, and a small loss of the target animals to severe weather. Restoration ecologists, on the other hand, try to achieve desired ecosystem dynamics by reducing or eliminating human perturbations and restoring natural ecosystem processes and the ecosystem components that drive these processes. Given the inherently dynamic state of most ecosystems, Boyce (1991, 1998) and others have suggested that a more appropriate management goal for YNP is to follow the "restoration" approach and maintain or restore ecological processes rather than try to maintain a particular ecological state. Management for processes would include maintaining or restoring the spatial and temporal variation that characterizes the natural ecosystem. Holling and Meffe (1996) persuasively argued that maintenance of natural variation is critical to the functioning of ecosystems and runs counter to most traditional management prescriptions. Because Yellowstone is influenced by periodic major events, both natural and human caused, it is probably impossible to maintain a particular state by active intervention. For example, the fires of 1988 resulted in substantial changes in the mosaic of vegetation communities, but these changes appear to be an integral component of the system and within the bounds of disturbances that periodically occurred in YNP (Romme and Despain 1989). Also, 1996 and 1997 floods throughout the GYE altered riparian communities and triggered new riparian recruitment, as expected from low-frequency, high-magnitude hydrological events (Skidmore et al. 1999).

If natural processes in YNP are to be managed or restored, we must change our focus from an emphasis on specific outcomes (the presence or absence of a species or state) to an emphasis on rates and variation. Ecological processes include production of crowd-pleasing cohorts of elk and bison calves in spring, but they also include the interrelationships between all species, including competition, predation, winter starvation, and changes in vegetation communities. Because ecological processes are dynamic, ecological communities change in time and space, with or without human intervention.

The need to understand and permit the full range of ecological processes is emphasized by interactions between disparate elements of the northern range. Frank et al. (1998) compared the grassy rangeland of the northern range to the Serengeti ecosystem in Kenya and Tanzania, an area that supports a higher diversity of large herbivores than the northern range. Nonetheless, herbivores have a key role in altering the transformation of materials in the functioning of both systems. Nutrient turnover rates are high in herbivore-dominated systems (including Yellowstone), and these grassland systems have rapid cycling of nutrients driven by high harvesting rates by herbivores. Removal of herbivores would transform the system into one dominated by detritivores, with slower cycling of nutrients.

Hobbs (1996) identified two major challenges to fully integrating the role of ungulates into ecosystem science. First, we need to better integrate the behavior of animals into ecosystem models. Many of the links between ungulates and ecosystem processes are the result of choices made by individual animals, such as selection of feeding sites, choice of forage items, and migration in response to climate, food availability, and other external pressures (e.g., hunting). Decisions about selection of habitats, feeding patches, and diets occur at a variety of scales (Senft et al. 1987, Bailey et al. 1996) and they have a profound influence on patterns of interaction between herbivores and ecosystems. Second, we need to better understand the interactions between population dynamics of animals and plants and ecosystem processes. Few studies have examined large-scale responses of ecosystems, including the response of animal and plant populations, to changes in herbivore density. Yellowstone's northern range may offer us an unusual opportunity for such studies. If the recent past is an indication of the future, we can expect large fluctuations in herbivore density and thus in their influence on ecosystem processes such as recycling and redistribution of materials, and successional dynamics.

Management for ecosystem processes remains a challenge for the future, and currently is more a conceptual guide than a prescription for immediate action. In their plea for more enlightened management of large systems, Holling and Meffe (1996) noted the following:

> Our advice to 'retain critical types and ranges of natural variation' must remain for the present as a management goal to which to aspire, as a conceptual underpinning for management, rather than an operational dictum. In practice this translates to adopting a conservative approach to changing parameters of systems we understand poorly but that we wish to manage. It means that the default condition, unless clearly proven otherwise, should be retention of the natural state rather than manipulation of system components or dynamics. It argues for humility when managing large systems (Stanley 1995).

The northern range's natural state is a dynamic one. Retention of natural processes is as close as we can come to this recommendation.

Despite our inability to manage natural processes, general guidelines are emerging for designing programs to monitor and detect environmental trends, and this remains an area of intensive research and evaluation (e.g., Dixon et al. 1998 and accompanying papers). It will be a challenge for YNP to look to opportunities of the future, without forgetting lessons from the past.

Public education also is important. The National Park Service (NPS) should explain the importance of ecosystem processes, trophic level relationships among species, primary production, and nutrient cycling. Although emphasis on biodiversity is certainly justified, the role of the area's landscape, climate, and history in maintaining the biodiversity of the area and its dynamic nature should be explained. That implies a focus on the web of life and its complexity in the lands under NPS jurisdiction and the change over time that characterizes natural systems, rather than on preconceptions about "the balance of nature" or the desirability of having many large, "charismatic" animals visible. NPS would do well to consider YNP a natural laboratory for public education, increasing public appreciation with an enhanced understanding gained through a park visit.

LARGE-SCALE INTERACTIONS AND PATTERNS

A second recurrent theme in our report has been the importance of spatial scale. The northern range is an incomplete ecosystem for large herbivores that rely on heterogeneity in the distribution of foods that vary seasonally in abundance, quality, and availability. A large spatial extent provides reserve areas that may not be preferred during normal years but can be used during times of shortage. The importance of the heterogeneity that normally accompanies a large spatial extent was emphasized by Walker et al. (1987), who examined drought-caused mortality of ungulates in African reserves that varied in size from 442 to 19,000 km^2. Mortality was relatively low in the large reserves because animals expanded their normal range and used reserve areas that were far from normal water sources during droughts. Walker et al. (1987) concluded that culling was unnecessary if there was sufficient spatial heterogeneity to provide reserve forage. Similarly, during severe winters ungulates in the northern range use areas outside park boundaries. However, many of these key areas are no longer accessible because of human activity and habitat fragmentation.

A large spatial extent is also important to preserve key ecosystem processes in the face of disturbances that recur over periods of centuries and affect areas of tens to thousands of square kilometers. The large fires that burned much of YNP in 1988 are the most obvious example of such a phenomenon; other examples include the eruption of Mount St. Helens in Washington and major floods. These major events create patchiness in the environment, and they may provide for the simultaneous occurrence of a critical set of characteristics that permit, for example, the establishment of, or change in, plant communities (Coughenour 1991, Turner et al. 1997, Foster et al. 1998).

In addition to providing forage reserves for ungulates, a large spatial extent allows animals to spread out the effects of their consumption. In YNP, ungulates use some areas heavily but others only lightly. In spring and early summer, the ungulates follow the emergence and greening-up of actively growing, nutritious plants, grazing intensively in a limited area for a period, then moving on, allowing the plants to recover. Thus, spatiotemporal heterogeneity is key to maintaining nutritious forages over an extended period, and the sequential greening of vegetation provides the impetus for herbivores to move on and allow the plants time to recover. These interactions should permit long-term sustainability of the system; however, intensive long-term use during extreme winter conditions may not permit some communities of woody plants to persist.

WEATHER, WOLVES, AND ASPEN IN YELLOWSTONE

Major controversy focuses on the causes of the virtual absence of recruitment of tree-sized aspen on the northern range since 1920 (Romme et al. 1995, Ripple and Larsen 2000b). Not all the circumstances that permitted aspen to recruit before 1920 are known, but the most important factor currently preventing recruitment of tree-sized aspen is heavy browsing by elk. If browsing by elk were greatly reduced or eliminated for a long enough time, as seems to have happened during the market hunting period of the 1880s (Romme et al. 1995), recruitment of tree-sized aspen would be likely under the current climate. What circumstances that previously existed, but are no longer present, might have permitted recruitment? The most obvious is that elk did not use aspen for winter survival because they migrated to lower areas with alternative winter food sources. Another possibility is that a combination of severe winters and a healthy predator population greatly reduced elk numbers or their distribution.

Weather during the 1800s—the end of the Little Ice Age—was consistently cooler and wetter than that of the 1900s (Chapter 2). This factor alone could account for smaller elk populations wintering on the northern range. In addition, wolves were present during the 1800s and they likely influenced the density and distribution of elk. Ripple and Larsen (2000b) suggested that wolves played a key role in the recruitment of aspen. They reviewed evidence showing that wolves can limit herbivore population size, but more importantly, wolves modify the location and feeding behaviors of ungulates that feed on aspen, thereby leading to localized recruitment of tree-sized aspen. Ripple and Larsen's hypothesis can account for small-scale recruitment of aspen, and with the addition of severe weather it can also account for synchronized, large-scale episodes of aspen recruitment. For this to occur may require the simultaneous effects of weather, and predation.

Severe weather during the winter following the 1988 fires resulted in the death of about 25% of the northern range elk population (Singer et al. 1989). Similar events occurred throughout the 1900s, most recently in 1996-1997. Elk populations have been subjected to annual harvest outside YNP ever since its establishment, and late-season hunting was initiated in 1968. Thus, the condition of the animals and the winter range is likely to have been better since 1968 than if the population size had been regulated solely by natural factors, including competition for forage and starvation. When subjected to a severe winter, a population strongly regulated by food supply and with limited ability to migrate out of the area is likely to experience high rates of mortality, resulting in a population considerably smaller than its prewinter size. If the elk population declined to a small fraction of what the northern range could support, then predation by wolves, whose numbers appear to be largely independent of elk population density, could prevent rapid recovery. A low enough density of elk would allow some aspen to grow tall, and sustained predation by wolves and other predators could maintain the elk population at a low density for long enough to permit recruitment of tree-sized aspen. In this scenario, establishment of tree-sized aspen requires (a) the elk population to decline rapidly after it has achieved a size too large to be maintained by the food available, (b) migration to be restricted, (c) a severe winter that causes starvation, and (d) a vigorous predator population that can keep the elk population from rapidly recovering. These conditions have been absent from the northern range since at least the late 1800s, when most (but not all) of the present tree-sized aspen stands were formed. Such a scenario is not greatly different from that which explains recruitment of fir on Isle Royale (Post et al. 1999), an island where long-range moose migration is prevented. Wolves hunted moose more efficiently during winters with heavy snowfall, thereby depressing moose populations and releasing fir from heavy browsing.

Several types of interactions have been proposed to account for predator-prey systems in which predation can maintain low densities of prey, but food limitations prevail at high densities (e.g., Walker and Noy-Meir 1982, Sinclair 1989, Boutin 1992). In general, theory suggests that prey populations are kept at lower levels only until predator populations decline or food sources increase. If this situation were to occur on the northern range, aspen recruitment would be episodic and occur at unpredictable and infrequent intervals.

INDICATORS OF UNACCEPTABLE CHANGE

If YNP continues to follow a policy that permits the natural range of variation, it will need to monitor ecosystem attributes that might indicate unacceptable change. Research in the park is only decades old, but some insights into past conditions are provided by analyses of lake sediments, tree rings, pollen profiles, and floodplain sediment profiles. These analyses of long-term trends identify the dynamic processes that led to current conditions of the Yellowstone ecosystem, but the linkages between past and present processes in the northern range have not been clearly demonstrated by research. Modification of the Yellowstone ecosystem through reintroduction of wolves, expansion of wintering areas for ungulates north of the park, and continued implementation of external hunting of ungulates, in the context of a changing climate, creates a degree of complexity that makes projection of long-term conditions in the park and northern range difficult. The committee consequently recommends that a comprehensive, integrated program of research and monitoring be established to measure the consequences of current and future changes in the external and internal driving variables. This program should include continued studies of animal and plant populations and their interactions, studies of predator-prey relationships, and studies of changes in the behavior of ungulates and predators as the system adjusts to the reestablishment of wolves. Concurrent studies of riparian and aspen recruitment; sagebrush communities; stream fluvial geomorphic processes in relation to riparian vegetation dynamics; rain, snow, surface flows, and groundwater levels; and other ecosystem components are also needed.

UNDERSTANDING THE CONSEQUENCES OF
ALTERNATIVE MANAGEMENT APPROACHES

Resource managers at YNP use natural regulation as the management approach for the biota of the northern range for scientific reasons and to meet public expectations. In any natural resource management context, the selection of an approach is inevitably in part a value judgment. What is our intent? What do we, as a society, or other decision-making level, want from or for the resource? Although managers generally strive to design multiple-use management approaches, in fact there often is an underlying policy purpose. In Yellowstone, managers could manage the system primarily to facilitate visitor interaction with animals, for ecosystem diversity, for scenic values, or for a combination of those values. The current decision to use natural regulation as opposed to management that actively reduces ungulate populations and thus decreases grazing and browsing pressure on the northern range is based in part on science (i.e., the determination that ungulate populations are at sustainable levels given the productivity of the range's vegetation), and in part it is a value judgment (based on the goals humans have set for the system).

The committee was asked to evaluate NPS's natural regulation management approach. It was not asked or appropriately constituted to look in depth at alternative management approaches. But in studying the dynamics of ungulate-ecosystem interactions in the northern range of the Yellowstone region and researching the impacts of natural regulation on the ecosystem, the committee gained insights about other approaches and learned lessons about associated scientific advantages and disadvantages. These insights may help YNP resource managers plan future actions for the northern range, use adaptive management principles, learn from the information generated, and change management approaches as needed, as more information becomes available. Adaptive management requires clearly defined goals, and it is predicated on use of a scientifically sound, comprehensive, integrated research program and long-term monitoring to determine the successes and/or consequences of management decisions.

The following text explores the scientific lessons that might be learned from various management approaches, including natural regulation. The committee recognizes that NPS managers must balance many factors beyond science in its decision making, but we can assist that process by projecting some of the possible ecological consequences of those decisions.

Reduction of Elk and Bison Populations Within the Park

Although the committee concludes that the number of ungulates in the northern range is less than the number at which density-dependent factors would cause it to decline (Chapter 4), experience from population reductions conducted in the 1950s and 1960s and from elk density/vegetation response studies elsewhere in the Rockies supports the view that a smaller population might allow recovery of some plant communities now degraded or unable to establish new recruits (e.g., woody riparian species including willows, aspen, and sagebrush communities). The likelihood that ungulate populations will be less than they have been recently is greater now that wolves are present

Experimental management to reduce ungulate populations, especially elk, and perhaps bison, could test the hypothesis that lower densities of these animals would allow increased recruitment of tree-sized aspen, expansion of willow communities, and growth of sagebrush to large sizes. The most effective way to reduce elk numbers in YNP would be to shoot them, but doing so might be contrary to the desires and values of the public. Visitors would see fewer of them, and shooting is likely to arouse strong public reaction. In addition, reducing ungulate numbers at this time would confound our ability to understand the effects of wolf reintroduction on ungulates. Finally, there is concern that a reduced ungulate population might disrupt food availability for the several wolf packs (and other predators) that now have a satisfactory food base within the park and lead the wolves to seek a domestic food base outside the park.

Reduction of Elk and Bison Populations Outside the Park

To test the hypothesis that reduced ungulate populations might allow recovery of woody plant communities, resource managers might experimentally reduce populations outside the park by working with the multiagency Northern Range Coordinating Committee to increase hunter harvest. This approach might partially test the concept that reduced elk numbers can enhance conditions of several northern range ecosystems (e.g., aspen, riparian, and sagebrush communities). An indirect social effect might be benefits to the local economy through increased outfitter clientele. However, this management approach also might confound our ability to understand the effects of wolf reintroduction, and the key disadvantage of the approach is that hunting success cannot be assured because elk might remain within the park, even during severe weather.

Improve Opportunities for Increased Out-Migration

Because lower elevation winter range outside the park has been greatly reduced, YNP resource managers could work with other state and federal agencies and land owners adjacent to the park to add more lands at lower elevations for winter use by ungulates. Elk herds throughout the northern Rockies tend to migrate from high to lower elevations as winter develops; the intensity of winter conditions usually influences the distance they move. Although hunting pressure at the park boundary may reduce migration seasonally, lack of open migration routes and land available for foraging at lower elevations also may influence migration. Lack of low-elevation winter range may eventually create an elk population that does not migrate outside the park but uses only the in-park northern range and higher elevation summer ranges. Already there are non-migratory elk populations within inner basins of YNP.

Continued efforts to increase land available for elk winter range might reduce ungulate effects on ecosystems within YNP during harsh winters or permit a large ungulate herd to be sustained within the northern range area with less damage to woody vegetation. Increasing the amount of winter habitat available also might prevent the transition of some of the northern range herd from migratory to nonmigratory, a phenomenon that over time could have long-term effects on the conditions of the northern range. This approach has numerous social and economic implications beyond the scope of this scientific assessment. For example, lands north of YNP in the Paradise Valley of the Yellowstone River have been used for ranching for decades, and many areas are fenced. At the same time, the human population of the Paradise Valley is increasing rapidly, giving rise to increased boundary controls and diverse opinions about wildlife use of private property. Finally, national forest lands in the mountains bordering the valley already have elk, and these animals usually move to lower elevations in limited areas in the valley in winter.

Natural Regulation

YNP resource managers consider the northern range to be in acceptable condition and the role and numbers of ungulates and other wildlife appropriate for a national park, and the best available scientific evidence does not indicate that ungulate populations are irreversibly damaging the northern range (Chapter 4). In addition, several significant changes have been made in the northern range in recent years, including the reintroduction of wolves and expansion of the winter range outside the park; the long-term influence of these changes cannot yet be determined. Thus, YNP resource managers could continue to manage the northern range as they are now. That is, YNP managers would continue to let the populations of elk, bison, and other ungulates fluctuate without any direct (inside Yellowstone) controls, letting a combination of weather, wolves, range conditions, and external controls (e.g., outside-the-park hunting, land uses, and population reduction by state agencies, such as the Montana Department of Livestock's program for bison) influence the population numbers.

Experimentation with continued use of natural regulation within YNP, recognizing the many external influences, would test whether the elk population has reached a dynamic equilibrium since the low numbers of the 1950s and 1960s. It would also allow time to observe the influences of the addition of a top predator and more available winter range. It will require careful monitoring to obtain full value from the experiment and to detect potentially serious changes in the ecosystem before they become severe or even irreversible.

CONCLUSIONS

Animal Populations

Density-dependent and density-independent factors interact to regulate the elk and bison populations in the northern range. Responses of elk and bison to potential regulatory factors are different: bison tend to expand their range when their populations exceed roughly 2,500, whereas reproductive rates in elk decline when their populations exceed roughly 15,000. Despite the density-dependent factors that affect elk and bison, their populations have fluctuated for a variety of reasons, including variation in weather and because ungulates and their food do not always vary in a synchronized way. Without rigorous management intervention, and perhaps even with it, ungulate populations will continue to fluctuate.

The pronghorn population has fluctuated widely and has been declining recently. Adverse factors include coyote predation and hunting on private land outside the park. Pronghorn may be affected by competition with elk, mule deer, and bison during severe winters. Bighorn sheep also may be responding adversely to many of these same factors.

Wolves will affect the population dynamics of ungulates as well as those of other predators in YNP, as they do elsewhere. The nature and magnitude of the effects are not predictable at present, although it is likely that wolves will reduce elk numbers. They might increase the magnitude or frequency of elk population fluctuations and might cause changes in the behavior of ungulates, especially elk, including changes in areas where they forage and spend time. The effect of wolves on bison is likely to be less variable and dramatic than their effect on elk, their primary prey in YNP.

Ungulates and Vegetation

Tree-sized aspen have not been added to the population in the northern range since about 1920. Currently, herbivory by elk is high enough to prevent any such recruitment, and apparently it has been since 1920. Although there have been fluctuations in climate since then, none has been large enough or persistent enough to account for the failure of aspen recruitment. Two un-tested hypotheses, working independently or in conjunction, could explain past recruitment. One is that enough elk migrated out of the park in severe winters to greatly reduce browsing pressure on aspen. The other is that wolves, before their extirpation, affected the distribution and abundance of elk so that at least some recruitment of tree-sized aspen and willows occurred even when elk were moderately abundant. If the latter were the case, then the wolves recently reintroduced into Yellowstone, including those in the northern range, could promote the recruitment of adult aspen and willows.

All tree-sized aspen in the northern range are more than 80 years old, and in the absence of recruitment, they will die out. Species associated with aspen will likely decline as well. Elk also are reducing the size and areal coverage of willows. Not enough is known about groundwater fluctuations or the role of secondary chemicals in herbivory to determine whether they are also affecting willow abundance.

Plant architecture and areal coverage of sagebrush has decreased during recent decades through browsing by elk, pronghorn, bison, and mule deer. In addition, herbivory has altered community composition, size, and recruitment. The effects are more significant at lower than at higher elevations in the northern range.

The composition and productivity of grassland communities in the northern range show little change with increasing grazing intensity. Humans, however, have changed the grasslands substantially by introducing exotic grasses and by other actions, many of which began before thorough inventories were initiated. Although conifer forests are used by ungulates, ungulates have little effect on conifer distribution and recruitment except for localized hedging of young conifers invading shrub and grassland areas.

The summer range does not seem to be limiting to the ungulate populations. Densities are relatively low on the summer range because the animals are spread out over larger areas than during winter-range use. Ungulates apparently have little effect on summer range communities, with the exception of young aspen, which are severely browsed.

The Northern Range

The condition of the northern range is different today than when Europeans first arrived in the area. The committee judges that the changes are the result of the larger numbers of elk and bison in the area, combined with human development and possibly climatic variability. The committee concludes, based on the best available evidence, that no major ecosystem component is likely to be eliminated from the northern range in the near or intermediate term. Further, although we recognize that the current balance between ungulates and vegetation does not satisfy everyone—there are fewer aspen and willows than in some similar ecosystems elsewhere—the committee concludes that the northern range is not on the verge of crossing some ecological threshold beyond which conditions might be irreversible. The same is true of the region's sagebrush ecosystems, despite reductions in the number and size of plants in some lower-elevation areas.

Natural Regulation

True natural regulation (i.e., letting nature take its course with no human intervention) has not been possible for more than a century, nor is it likely to become possible in Yellowstone's foreseeable future. Because of development on the park's borders, ungulates do not have free access to areas outside YNP that they formerly used during times of environmentally imposed stress. Because ungulate populations are influenced by activities both inside and outside the park, the conclusions in this report should not be interpreted as either vindication or criticism of YNP's natural regulation policy.

YNP's practice of intervening as little as possible is as likely to lead to the maintenance of the northern range ecosystem and its major components as any other practice. If the park decides that it needs to intervene to enhance declining species like aspen, the smaller the intervention, the less likely it is to do unintended damage. For example, if YNP decided to maintain tree-sized aspen in the park, putting exclosures around some stands would be an intervention much less likely to trigger unanticipated processes than an attempt to eliminate or greatly reduce populations of ungulates.

Large ecosystems in general and YNP's northern range in particular are dynamic. They change in sometimes unpredictable ways. The recent reintroduction of wolves, which has restored an important component of this ecosystem, adds to the dynamism, complexity, and uncertainty, especially in the short term. The near future promises to be most instructive about how elk and other ungulates interact with a complete community of predators.

RECOMMENDATIONS

Given the complexities involved in managing Yellowstone's dynamic ecosystems, there is a continuing need for rigorous research and public education. The committee offers the following recommendations designed to enhance understanding of key processes affecting Yellowstone's ungulate populations, vegetation, and ecological processes.

Park Management and Interpretation

• To the degree possible, all management at YNP should be done as adaptive management. This means that actions should be designed to maximize their ability to generate useful, scientifically defensible information, including quantitative models, and that the results of actions must be adequately monitored and interpreted to provide information about their consequences to guide subsequent actions.

• There is insufficient scientific knowledge available to enable us to predict the consequences of different management approaches. Thus, long-term scientific investigations and experiments are needed to provide solid scientific evidence for evaluating management options.

• The NPS educational and outreach program can play an important role in fostering public understanding of the complex and dynamic nature of ungulate ecology in the GYE, which is an essential adjunct to effective management of northern Yellowstone ungulates. Therefore, we encourage the NPS to increase its focus on entire ecosystem relationships, processes, and dynamics of the GYE, especially emphasizing the importance of primary production and trophic-level relationships.

Vegetation

• A rigorous study focusing on aspen populations throughout the GYE should be undertaken to quantify the relative importance of the factors known or hypothesized to influence aspen stand structure. It should include establishing an increased number of large exclosures with a long-term commitment to monitoring the effects of restricting herbivory by ungulates. The study sites should be discussed in the NPS ecosystem interpretive program.

• A careful examination of the variables that are most strongly affecting the riparian ecosystems on the northern range is needed, especially the relationship between herbivory and groundwater availability. This should include an understanding of fluvial processes, surface and groundwater hydrology, and biotic processes.

• Research should continue on northern range sagebrush/grassland communities.

• Research is needed to determine whether it is possible to differentiate ungulate use of tall and short willows based on both the food-deprivation levels of the ungulates (i.e., winter starvation) and the levels of secondary chemicals in the plants.

Animal Populations

• The behavioral adaptations of elk and other ungulates, and the changes in patterns of habitat use as a consequence of the presence of the wolf as a large predator newly restored to the system, should be closely monitored as a basis for understanding the dynamic changes that are taking place within the system.

• The changes taking place in the interactions among the large predators of YNP and their effects on the trophic dynamics of the ecosystem should be closely monitored as wolves become an established component of the system.

• A thorough study of current and likely future trajectories of the pronghorn population and the role of human effects on this population is needed, including the influence of disturbance by visitors and the Stevens Creek bison facility. The study should evaluate the likely consequences of a full range of potential management options, from doing nothing to actively controlling predators and providing artificial winter feed.

• Periodic surveillance for pathogens (including brucellosis) in wild ruminants in the northern range should be continued, and a more thorough understanding of population-level threshold dynamics gained. Samples could routinely be obtained from animals immobilized for research, found dead, or killed by hunters.

Biodiversity

• A periodic (every 10-15 years) and comprehensive biodiversity assessment is needed on the northern range to evaluate potential direct and indirect impacts of ungulate grazing, both of terrestrial and aquatic environments. Initially, species should be identified as consistent indicators of habitat change. These species should then be monitored intensively during periods between comprehensive assessments.

Human Influences

- A comprehensive research effort is needed to assess the influence of seasonal densities, distribution, movements, and activities of people within YNP and adjacent areas on wildlife species, their habitat use patterns, behavior, foraging efficiency, vegetation impacts, and other aspects of their ecosystem relationships.
- The effects of changing land-use patterns in the landscape surrounding Yellowstone on the park's biota and natural processes, such as fire, need to be investigated.

EPILOGUE

GYE is dynamic, and change is a normal part of the system as far back as we have records or can determine from physical evidence. Based on that record of change, it is certain that sooner or later the environment of the GYE will change in ways that cause the loss of some species and changes in community structure. Human-induced changes, both within the GYE and globally, are likely to accelerate these changes.

Although dramatic ecological change does not appear to be imminent, it is not too soon for the managers of YNP and others to start thinking about how to deal with potential changes. Before humans modified the landscape of the GYE—limiting access to much of it and interrupting migration routes—animals could respond to environmental changes by moving to alternative locations. To a lesser degree, and over longer time frames, plants could adapt as well, especially in places with significant topographic relief. But many options that organisms formerly had for dealing with environmental changes have been foreclosed because of human development of the region. Human-induced climate change is expected to be yet another long-term influence on the ecosystem. Reconciling the laudable goals of preserving ecosystem processes and associated ecosystem components with human interests and influences on wildlands will be a growing challenge in the future, not only in the GYE. That reconciliation will involve conflicting policy goals, incomplete scientific information, and management challenges. Resolving these conflicts will require all the vision, intellectual capacity, financial resources, and goodwill that can be brought to bear on them.

References

Allen, E.O. 1971. White-tailed deer. Pp. 69-79 in Game Management in Montana, T.W. Mussehl, and F.W. Howell, eds. Federal Aid Project W-3-C. Montana Fish and Game Department, Helena.

Allen-Diaz, B., and J.W. Bartolome. 1998. Sagebrush-grass vegetation dynamics: Comparing classical and state-transition models. Ecol. Appl. 8(3):795-804.

Alley, R.B. 2000. Ice-core evidence of abrupt climate changes. Proc. Natl. Acad. Sci. USA 97(4):1331-1334.

Alley, R.B., P.A. Mayewski, T. Sowers, M. Stuiver, K.C. Taylor, and P.U. Clark. 1996. Holocene Climatic Instability: A Large, Widespread Event 8,200 Years Ago. 1996 Fall Meeting of the American Geophysical Union, San Francisco, CA.

Anders, M.H., J.W. Geissman, L.A. Piety, and J.T. Sullivan. 1989. Parabolic distribution of Circumeastern Snake River plain seismicity and latest Quaternary faulting: migratory pattern and association with the Yellowstone hotspot. J. Geophys. Res. B 94(2):1589-1621.

Anderson, J. 1991. A conceptual framework for evaluating and quantifying naturalness. Conserv. Biol. 5(3):347-352.

Archer, S. 1994. Woody plant encroachment into southwestern grasslands and savannas: Rates, patterns and proximate causes. Pp. 130-138 in Ecological Implications of Livestock Herbivory in the West, M. Vavra, W.A. Laycock, and R.D. Pieper, eds. Denver, CO: Society for Range Management.

Arno, S.F. 1980. Forest fire history in the northern Rockies. J. For. 78(8):460-465.

Austin, P.J., L.A. Suchar, C.T. Robbins, and A.E. Hagerman. 1989. Tannin-binding proteins in saliva of deer and their absence in saliva of sheep and cattle. J. Chem. Ecol. 15(4):1335-1347.

Bailey, D.W., J.E. Gross, E.A. Laca, L.R. Rittenhouse, M.B. Coughenour, D.M. Swift, and P.L. Simms. 1996. Mechanisms that result in large herbivore grazing distribution patterns. J. Range Manage. 49(5):386-400.

Bailey, V. 1930. Animal Life of Yellowstone National Park. Springfield, IL: Charles C Thomas. 241 pp.

Baker, W.L., J.A. Munroe, and A.E. Hessl. 1997. The effects of elk on aspen in the winter range of Rocky Mountain National Park. Ecography 20(2):155-165.

Balling, R.C., Jr., G.A. Meyer, and S.G. Wells. 1992a. Climate change in Yellowstone National Park: is the drought-related risk of wildfires increasing? Clim. Change 22(1):35-45.

Balling, R.C., Jr., G.A. Meyer, and S.G. Wells. 1992b. Relation of surface climate and burned area in Yellowstone National Park. Agric. For. Meteorol. 60(3/4):285-293.

Bangs, E.E., S.H. Fritts, J.A. Fontaine, D.W. Smith, K.M. Murphy, C.M. Mack, and C.C. Niemeyer. 1998. Status of gray wolf restoration in Montana, Idaho, and Wyoming. Wildl. Soc. Bull. 26(4):785-798.

Barmore, W.J., Jr., 1965. Aspen-Elk Relationships on the Northern Yellowstone Winter Range. Paper presented at the Western Association of State Fish and Game Commissions Elk Workshop, Bozeman, MT, March 2-4, 1965. 16 pp.

Barmore, W.J., Jr. 1980. Population Characteristics, Distribution, and Habitat Relationships of Six Ungulate Species on Winter Range in Yellowstone National Park. Final Report, Yellowstone National Park, Wyoming. 677 pp.

Barnes, B.V. 1966. The clonal growth habit of American aspens. Ecology 47(3):439-447.

Barnes, V.G., and O.E. Bray. 1967. Population Characteristics and Activities of Black Bears in Yellowstone National Park. Final report. Colorado Cooperative Wildlife Research Unit, Colorado State University, Fort Collins, CO.

Barnosky, C.W., P.M. Anderson, and P.J. Bartlein. 1987. The northwestern U.S. during deglaciation: Wegetational history and paleoclimatic implications. Pp. 289-321 in North America and Adjacent Oceans During the Last Deglaciation, The Geology of North America, Vol. K-3, W.F. Ruddiman, and H.E. Wright, Jr., eds. Boulder, CO: Society of America.

Barnosky, C.W., H.E. Wright, Jr., D.R. Engstrom, and S.C. Fritz. 1988. The relationship of climate to sedimentation rates in lakes and ponds. Pp. 4 in First Annual Meeting of Research and Monitoring on Yellowstone's Northern Range, Mammoth, WY, January 28-29, 1988, F.J. Singer, ed. Yellowstone National Park, WY: U.S. Department of the Interior, National Park Service.

Barnosky, E.H. 1994. Ecosystem dynamics through the past 2000 years as revealed by fossil mammals from Lamar Cave in Yellowstone National Park, USA. Hist. Biol. 8(1/4):71-90.

Barrett, M.W. 1982. Distribution, behavior, and mortality of pronghorns during a severe winter in Alberta, Canada. J. Wildl. Manage. 46(4):991-1002.

Barrett, S.W. 1994. Fire regimes on andesitic mountain terrain in northeastern Yellowstone National Park, Wyoming. Int. J. Wildland Fire 4(2):65-76.

Bartlein, P.J., C. Whitlock, and S.L. Shafer. 1997. Future climate in the Yellowstone

National Park region and its potential impact on vegetation. Conserv. Biol. 11(3):782-792.

Bartos, D.L., and W.F. Mueggler. 1981. Early succession in aspen communities following fire in western Wyoming. J. Range Manage. 34(4):315-318.

Bartos, D.L., J.K. Brown, and G.D. Booth. 1994. Twelve years biomass response in aspen communities following fire. J. Range Manage. 47(1):79-83.

Bayless, S.R. 1969. Winter food habitats, range use, and home range of antelope in Montana. J. Wildl. Manage. 33(3):538-551.

Beetle, A.A. 1974. Range Survey in Teton County, Wyoming: Part IV. Quaking Aspen. Science Monograph 27. Agricultural Experiment Station, University of Wyoming, Laramie, WY. 28 pp.

Beetle, A.A. 1979. Jackson Hole elk herd: A summary after 25 years of study. Pp. 259-262 in North American Elk: Ecology, Behavior, and Management, M.S. Boyce, and L.D. Hayden-Wing, eds. University of Wyoming, Laramie, WY.

Begon, M., M. Mortimer, and D.J. Thompson. 1996. Population Ecology: A Unified Study of Animals and Plants, 3rd Ed. Cambridge, MA: Blackwell Science.

Belovsky, G.E. 1981. Food plant selection by a generalist herbivore: the moose. Ecology 62(4):1020-1030.

Berger, J., J.E. Swenson, and I.L. Persson. 2001. Recolonizing carnivores and naïve prey: conservation lessons from Pleistocene extinctions. Science 291(5506):1036-1039.

Bessie, W.C., and E.A. Johnson. 1995. The relative importance of fuels and weather on fire behavior in subalpine forests. Ecology 76(3):747-762.

Bjornstad, O.N., J.M. Fromentin, N.C. Stenseth, and J. Gjosaeter. 1999. A new test for density-dependent survival: the case of coastal cod populations. Ecology 80(4):1278-1288.

Boutin, S. 1992. Predation and moose population dynamics: a critique. J. Wildl. Manage. 56(1):116-127.

Boyce, M.S. 1989. The Jackson Elk Herd: Intensive Wildlife Management in North America. New York: Cambridge University Press.

Boyce, M.S. 1991. Natural regulation or the control of nature? Pp. 183-208 in The Greater Yellowstone Ecosystem: Redefining America's Wilderness Heritage, R.B. Keiter, and M.S. Boyce, eds. New Haven: Yale University Press.

Boyce, M.S. 1998. Ecological-process management and ungulates: Yellowstone's conservation paradigm. Wildl. Soc. Bull. 26(3):391-398.

Boyce, M.S., and E.M. Anderson. 1999. Evaluating the role of carnivores in the Greater Yellowstone ecosystem. Pp. 265-284 in Carnivores in Ecosystems: The Yellowstone Experience, T.W. Clark, A.P. Curlee, S.C. Minta, and P.M. Kareiva, eds. New Haven, CT: Yale University Press.

Boyce, M.S., and J.M. Gaillard. 1992. Wolves in Yellowstone, Jackson Hole, and the North Fork of the Shoshone River: Simulating ungulate consequences of wolf recovery. Pp. 4.71-4.115 in Wolves for Yellowstone? A Report to the United

States Congress, Vol. 4. Research and Analysis, J.D. Varley, and W.G. Brewster, eds. National Park Service, Yellowstone National Park. July.

Boyce, M.S., and E.H. Merrill. 1991. Effects of the 1988 fires on ungulates in Yellowstone National Park. Proceedings of the Tall Timbers Fire Ecology Conference 17:121-132.

Boyce, M.S., B.M. Blanchard, R.R. Knight, and C. Servheen. 2001. Population Viability for Grizzly Bears: A Critical Review. IBA Monograph No. 4. International Association for Bear Research and Management, IBA, University of Tennessee, Knoxville, TN. [Online]. Available: http://www.bearbiology.com [Nov.12, 2001].

Bradley, R. 2000. Enhanced: 1000 years of climate change. Science 288(5470):1353-1355.

Brain, C.K. 1981. The Hunters or the Hunted? An Introduction to African Cave Taphonomy. Chicago: University of Chicago Press.

Brichta, P.H. 1987. Environmental Relationships Among Wetland Community Types of the Northern Range, Yellowstone National Park. M.S. Thesis, University of Montana, Missoula, MT. 74 pp.

Brown, J.K., and N.V. DeByle. 1987. Fire damage, mortality, and suckering in aspen. Can. J. For. Res. 17:1100-1109.

Bruns, E.H. 1970. Winter predation of golden eagles and coyotes on pronghorn antelopes. Can. Field Nat. 84:301-304.

Bryant, J.P., T.P. Clausen, P.B. Reichardt, M.C. McCarthy, and R.A. Werner. 1987. Effects of nitrogen fertilization upon the secondary chemistry and nutritional value of quaking aspen (Populus tremuloides Michx.) leaves for the large aspen torix (Choristoneura conflictama (Walker). Oecologia 73(4):513-517.

Bunch, T.D., W.M. Boyce, C.P. Hibler, W.R. Lance, T.R. Spraker, and E.S. Williams. 1999. Diseases of North American wild sheep. Pp. 209-237 in Mountain Sheep of North America, R. Valdez, and P.R. Krausman, eds. Tuscon, AZ: University of Arizona Press.

Busch, D.E., N.L. Ingraham, and S.D. Smith. 1992. Water uptake in woody riparian phreatophytes of the southwestern United States: a stable isotope study. Ecol. Appl. 2(4):450-459.

Buskirk, S.W. 1999. Mesocarnivores of Yellowstone. Pp. 165-187 in Carnivores in Ecosystems: The Yellowstone Experience, T.W. Clark, A.P. Curlee, S.C. Minta, and P.M. Kareiva, eds. New Haven, CT: Yale University Press.

Byers, J.A. 1997. American Pronghorn: Social Adaptations and the Ghost of Predators Past. Chicago: University of Chicago Press.

Byers, J.A., and J.D. Moodie. 1990. Sex-specific maternal investment in pronghorn, and the question of a limit on differential provisioning in ungulates. Behav. Ecol. Sociobiol. 26(3):157-164.

Callicott, J.B. 1982. Traditional American Indian and western European attitudes toward nature: An overview. Environ. Ethics 4:293-318.

Cannon, K.P. 1992. A review of archeological and paleontological evidence for the prehistoric presence of wolf and related prey species in the Northern and Central Rockies physiographic provinces. Pp. 1.175-1.266 in Wolves for Yellowstone? A Report to the United States Congress, Vol. IV: Research and Analysis, J.D. Varley and W.G. Brewster, eds. National Park Service, Yellowstone National Park. July.

Cannon, K.P. 1997. A Review of Prehistoric Faunal Remains from Various Contexts in Yellowstone National Park, Wyoming. U.S. Department of the Interior, National Park Service, Midwest Archeological Center, Lincoln, NE.

Carbyn, L.N. 1987. Gray wolf and red wolf. Pp. 358-376 in Wild Furbearer Management and Conservation in North America, M. Novak, J.A. Baker, M.E. Obbard, and B. Malloch, eds. Ontario: Ministry of Natural Resources.

Caslick, J.W. 1998. Yellowstone pronghorns: relict herd in a shrinking habitat. Yellowstone Sci. 6(4):20-24.

Caslick, J., and E. Caslick. 1999. Pronghorn Distribution in Winter 1998-1999 Yellowstone National Park. YCR-NR-99-2. Yellowstone National Park, Wyoming. March 24, 1999.

Chadde, S.W., and C.E. Kay. 1991. Tall-willow communities on Yellowstone's Northern Range: A test of the "Natural Regulation" paradigm. Pp. 231-262 in The Greater Yellowstone Ecosystem: Redefining America's Wilderness Heritage, R.B. Keiter and M.S. Boyce, eds. New Haven: Yale University Press.

Chadde, S.W., P.L. Hansen, and R.D. Pfister. 1988. Wetland Plant Communities of the Northern Range, Yellowstone National Park. School of Forestry, University of Montana, Missoula, MT. 81 pp.

Chase, A. 1986. Playing God in Yellowstone: The Destruction of America's First National Park. Boston: Atlantic Monthly Press.

Chase, J.D., W.E. Howard, and J.T. Roseberry. 1982. Pocket gophers (Geomyidae). Pp. 239-255 in Wild Mammals of North America: Biology, Management, and Economics, J.A. Chapman, and G.A. Feldhamer, eds. Baltimore: Johns Hopkins University Press.

Chomko, S.A., and B.M. Gilbert. 1987. The late Pleistocene/Holocene faunal record in the Northern Bighorn Mountains, Wyoming. Pp. 394-408 in Late Quaternary Mammalian Biogeography and Environments of the Great Plains and Prairies, R.W. Graham, H.A. Semken, Jr., and M.A. Graham, eds. Scientific Papers Vol. 22. Springfield, IL: Illinois State Museum.

Christensen, N.L., J.K. Agee, P.F. Brussard, J. Hughes, D.H. Knight, G.W. Minshall, J.M. Peek, S.J. Pyne, F.J. Swanson, J.W. Thomas, S. Wells, S.E. Williams, and H.A. Wright. 1989. Interpreting the Yellowstone fires of 1988. BioScience 39(10):678-685.

Clark, T.W., S.C. Minta, P. Curlee, and P.M. Kareiva. 1999. A model ecosystem for carnivores in Greater Yellowstone. Pp. 1-9 in Carnivores in Ecosystems: The Yellowstone Experience, T.W. Clark, A.P. Curlee, S.C. Minta, and P.M. Kareiva, eds. New Haven, CT: Yale University Press.

Cole, G.F. 1969. The Elk of Grand Teton and Southern Yellowstone National Parks.

Research Report GRTE-N-1. U.S. National Park Service, Office of Natural Science Studies, Washington, DC. 80 pp.

Cole, G.F. 1971. An ecological rationale for the natural or artificial regulation of native ungulates in parks. N. Amer. Wildl. Natur. Resour. Conf. Trans. 36:417-425.

Cole, G.F. 1976. Management Involving Grizzly and Black Bears in Yellowstone National Park, 1970-1975. Natural Resources Report No. 9. National Park Service, U.S. Department of the Interior, Washington, DC.

Cook, R.E. 1983. Clonal plant populations. Am. Sci. 71(3):244-253.

Cooper, H.W. 1963. Range Site and Condition Survey Northern Elk Range Yellowstone Park June 3-14, 1963. USDA Soil Conservation Service, Yellowstone National Park. 9 pp.

Coppock, D.L., and J.K. Detling. 1986. Alteration of bison and black-tailed prairie dog grazing interaction by prescribed burning. J. Wildl. Manage. 50(3):452-455.

Coughenour, M.B. 1991. Biomass and nitrogen responses to grazing or upland steppe on Yellowstone's northern winter range. J. Appl. Ecol. 28(1):71-82.

Coughenour, M.B., and F.J. Singer. 1991. The concept of overgrazing and its application to Yellowstone's Northern Range. Pp. 209-230 in The Greater Yellowstone Ecosystem: Redefining America's Wilderness Heritage, R.B. Keiter, and M.S. Boyce, eds. New Haven: Yale University Press.

Coughenour, M.B., and F.J. Singer. 1996a. Elk population processes in Yellowstone National Park under the policy of natural regulation. Ecol. Appl. 6(2):573-593.

Coughenour, M.B., and F.J. Singer. 1996b. Yellowstone elk population responses to fire-a comparison of landscape carrying capacity and spatial-dynamic ecosystem modeling approaches. Pp. 169-180 in The Ecological Implications of Fire in Greater Yellowstone: Proceedings Second Biennial Conference on the Greater Yellowstone Ecosystem, Yellowstone National Park, September 19-21, 1993, J.M. Greenlee, ed. Fairfield, WA: International Association of Wildland Fire.

Crabtree, R.L., and J.W. Sheldon. 1999a. Coyotes and canid coexistence in Yellowstone. Pp. 127-164 in Carnivores in Ecosystems: The Yellowstone Experience, T.W. Clark, A.P. Curlee, S.C. Minta, and P.M. Kareiva, eds. New Haven, CT: Yale University Press.

Crabtree, R.L., and J.W. Sheldon. 1999b. The ecological role of coyotes on Yellowstone's northern range. Yellowstone Sci. 7(2):15-23.

Craighead, F.L., M.E. Gilpin, and E.R. Vyse. 1999. Genetic considerations for carnivore conservation in the Greater Yellowstone ecosystem. Pp. 285-322 in Carnivores in Ecosystems: The Yellowstone Experience, T.W. Clark, A.P. Curlee, S.C. Minta, and P.M. Kareiva, eds. New Haven: Yale University Press.

Creasman, S.D., T.P. Reust, J.C. Newberry, K. Harvey, J.C. Mackey, C. Moore, D. Kullen, and I. Pennella. 1982. Archaeological Investigations Along the Trailblazer Pipeline. Cultural Resources Management Report 3. Rock Springs, WY: Western Wyoming College.

Crowley, T.J., and G.R. North. 1991. Paleoclimatology. New York: Oxford University Press.

Curtis, J.T. 1955. A prairie continuum in Wisconsin. Ecology 36(4):558-566.

Dale, B.W., L.G. Adams, and R.T. Bowyer. 1995. Winter wolf predation in a multiple prey system, Gates of the Arctic National Park, Alaska. Pp. 223-230 in Ecology and Conservation of Wolves in a Changing World, L.N. Carbyn, S.H. Fritts, and D.R. Seip, eds. Edmonton, Alberta: Canadian Circumpolar Institute.

Dansgaard, W., S.J. Johnsen, H.B. Clausen, D. Dahl-Jensen, and N.S. Gunderstrup. 1993. Evidence for general instability of past climate from a 250-kyr ice-core record. Nature 364(6434):218-220.

Daubenmire, R.F. 1943. Vegetation zonation in the Rocky Mountains. Bot. Rev. 9(6):325-393.

Davis, M.B. 1976. Pleistocene biogeography of temperate deciduous forests. Geoscience and Man 13:13-26.

Dawson, T.E., and J.R. Ehleringer. 1991. Streamside trees that do not use stream water. Nature 350(6316):335-337.

DeByle, N.V., and R.P. Winokur, eds. 1985. Aspen: Ecology and Management in the Western United States. Gen. Tech. Rep. RM-119. Fort Collins, CO: U.S. Department of Agriculture Forest Service, Rocky Mountain Forest and Range Experiment Station.

Dekker, D., W. Bradford, and J. R. Gunson. 1995. Elk and wolves in Jasper National Park, Alberta, from historical times to 1992. Pp. 85-94 in Ecology and Conservation of Wolves in a Changing World, L.N. Carbyn, S.H. Fritts, and D.R. Seip, eds. Edmonton, Alberta: Canadian Circumpolar Institute.

DelGiudice, G.D, F.J. Singer, and U.S. Seal. 1991. Physiological assessment of winter nutritional deprivation of elk in Yellowstone National Park. J. Wildl. Manage. 55(4):653-664.

Dennis, B., and M.L. Taper. 1994. Density dependence in time series observations of natural populations: estimation and testing. Ecol. Monogr. 64(2):205-224.

Despain, D.G. 1990. Yellowstone Vegetation: Consequences of Environment and History in a Natural Setting. Boulder: Roberts Rinehart. 239 pp.

Despain, D.G. 1996. The coincidence of elk migration and flowering of bluebunch wheatgrass. Pp. 139-143 in Effects of Grazing by Wild Ungulates in Yellowstone National Park, F.J. Singer, ed. Tech. Rep. NPS/NRYELL/NRTR/96-01. U.S. Department of the Interior, National Park Service, Natural Resource Information Division, Denver, CO.

Despain, D., A. Rodman, P. Schullery and H. Shovic. 1989. Burned Area Survey of Yellowstone National Park: The Fires of 1988. Division of Research and Geographic Information Systems Laboratory, Yellowstone National Park, WY.

Dillehay, T.D. 1974. Late Quaternary Bison population changes on the southern Plains. Plains Anthropologist 19:180-196.

Dixon, P.M., A.R. Olsen, and B.M. Kahn. 1998. Measuring trends in ecological resources. Ecol. Appl. 8(2):225-227.

Dobson, A., and M. Meagher. 1996. The population dynamics of brucellosis in the Yellowstone National Park. Ecology 77(4):1026-1036.

Dyksterhuis, E.J. 1949. Condition and management of rangeland based on quantitative ecology. J. Range Manage. 2:104-115.

Eberhardt, L.L. 1995. Population trend estimates from reproductive and survival rates. Pp. 13-19 in Yellowstone Grizzly Bear Investigations, Annual Report of the Interagency Study Team, 1994. Bozeman, MT: National Biological Service.

Eberhardt, L.L., and R.R. Knight. 1996. How many grizzlies in Yellowstone? J. Wildl. Manage. 60(2):416-421.

Eberhardt, L.L., B.M. Blancard, and R.R. Knight. 1994. Population trend of the Yellowstone grizzly bears as estimated from reproductive and survival rates. Can. J. Zool. 72(2):360-363.

Eckenwalder, J.E. 1996. Systematics and evolution of Populus. Pp. 7-32 in Biology of Populus and Its Implications for Management and Conservation, R.F. Stettler, H.D. Bradshaw, Jr., P.E. Heilman, and T.M. Hinckley, eds. Ottawa: NRC Research Press, National Research Council of Canada.

Einspahr, D.W., and L.L. Winston. 1977. Genetics of Quaking Aspen. Research Paper WO 25. Washington, DC: U.S. Department of Agriculture, Forest Service.

Elliot, C.R., and M.M. Hektner. 2000. Wetland Resources of Yellowstone National Park. Yellowstone National Park, WY. 32 pp. [Online]. Available: http://www.nps.gov/yell/publications/pdfs/wetlands/index.htm. [October 24, 2001].

Engstrom, D.R., C. Whitlock, S.C. Fritz, and H.E. Wright, Jr. 1991. Recent environmental changes inferred from the sediments of small lakes in Yellowstone's northern range. J. Paleolimnol. 5:139-174.

Erb, J.D., and M.S. Boyce. 1999. Distribution of population declines in large mammals. Conserv. Biol. 13(1):199-201.

Erwin, E.A., M.G. Turner, R.L. Lindroth, and W.H. Romme. 2001. Secondary plant compounds in seedling and mature aspen (Populus tremuloides) in Yellowstone National Park, Wyoming. Am. Midland Nat. 145(2):299-308.

Farnes, P.E. 1996. An index of winter severity for elk. Pp. 303-306 in Effects of Grazing by Wild Ungulates in Yellowstone National Park, F.J. Singer, ed. Tech. Rep. NPS/NRYELL/NRTR/96-01. U.S. Department of the Interior, National Park Service, Natural Resource Information Division, Denver, CO.

Farnes, P.E. 1998. Water yield increase and climatic variability after the 1988 fires in the Yellowstone and Madison River drainages. Pp. 34 in Making a Place for Nature, Seeking Our Place in Nature, 125th Anniversary Symposium, Yellowstone National Park, Montana State University, Bozeman, MT, May 11-23, 1998, S.L. Consolo Murphy, ed. Yellowstone National Park, WY: Yellowstone Association.

Farnes, P.E., C. Heydon, and K. Hansen. 1999. Snowpack Distribution Across Yellowstone National Park, Wyoming. Final Report. Contract No. 291601. Yellowstone Center for Resources, Yellowstone National Park, WY.

Faunmap Working Group. 1994. FAUNMAP: A Database Documenting Late Quaternary Distributions of Mammal Species in the United States, R.W. Graham, and E.L. Lundelius, Jr., codirectors. Scientific Papers 25. Springfield, IL: Illinois State Museum.

Faunmap Working Group (R.W. Graham, E.L. Lundelius, Jr., M.A. Graham, E.K. Schroeder, R.S. Toomey, III, E. Anderson, A.D. Barnosky, J.A. Burns, C.S. Churcher, D.K. Grayson, R.D. Guthrie, C.R. Harington, G.T. Jefferson, L.D.Martin, H.G. Mcdonald, R.E. Morlan, H.A. Semken, Jr., S.D. Webb, L. Werdelin, and M.C. Wilson). 1996. Spatial response of mammals to late quaternary environmental fluctuations. Science 272(5268):1601-1606.

Flanagan, L.B., J.R. Ehleringer, and T.E. Dawson. 1992. Water sources of plants growing in woodland, desert, and riparian communities: evidence from stable isotope analysis. Pp. 43-47 in Proceedings- Symposium of Ecology and Management of Riparian Shrub Communities, Sun Valley, ID, May 29-31, 1991, W.P. Clary, E.D. McArthur, D. Bedunah, and C.L. Wambolt, eds. Gen. Tech. Report INT-289. Ogden, UT: Intermountain Research Station, Forest Service, U.S. Department of Agriculture.

Flannigan, M.D., and C.E. Van Wagner. 1991. Climate change and wildfire in Canada. Can. J. For. Res. 21(1):66-72.

Foster, D.R., D.H. Knight, and J.F. Franklin. 1998. Landscape patterns and legacies resulting from large infrequent forest disturbances. Ecosystems 1(6):497-510.

Fowells, H.A. 1965. Silvics of Forest Trees of the United States. Agric. Handb. No. 271. Washington, DC: Division of Timber Management Research, Forest Service.

Fowler, C.W. 1981. Density dependence as related to life history strategy. Ecology 62(3):602-610.

Fowler, C.W. 1987. A review of density dependence in populations of large mammals. Pp. 401-441 in Current Mammalogy, Vol. 1., H.H. Genoways, ed. New York: Plenum Press.

Frank, D.A., and P.M. Groffman. 1998. Ungulate vs. landscape control of soil C and N processes in grasslands of Yellowstone National Park. Ecology 79(7):2229-2241.

Frank, D.A., and S.J. McNaughton. 1992. The ecology of plants, large mammalian herbivores, and drought in Yellowstone National Park. Ecology 73(6):2043-2058.

Frank, D.A., S.J. McNaughton, and B.F. Tracy. 1998. The ecology of the Earth's grazing ecosystems. Bioscience 48(7):513-521.

Friedman, J.M., M.L. Scott, and W.L. Lewis, Jr. 1995. Restoration of riparian forest using irrigation, artificial disturbance, and natural seedfall. Environ. Manage. 19(4):547-557.

Frison, G.C. 1974. The Casper Site: A Hell Gap Bison Kill on the High Plains. New York: Academic Press.

Frison, G.C. 1978. Prehistoric Hunters of the High Plains. New York: Academic Press.

Frison, G.C., and B.A. Bradley. 1991. Prehistoric Hunters of the High Plains. New York: Academic Press.

Frison, G.C., and D.J. Stanford. 1982. The Agate Basin Site: A Record of the Paleoindian Occupation of the Northwestern High Plains. New York: Academic Press.

Gaillard, J.M., M. Festa-Bianchet, and N.G. Yoccoz. 1998. Population dynamics of large herbivores: variable recruitment with constant adult survival. Trends Ecol. Evol. 13(2):58-63.

Gleason, H.A. 1926. The individualistic concept of the plant association. Bull. Torrey Bot. Club 53(1):7-26.

Glendening, G.E. 1952. Some quantitative data on the increase of mesquite and cactus on a desert grassland range in southern Arizona. Ecology 33(3):319-328.

Gogan, P.J., E.M. Olexa, T.O. Lemke, and K. Podruzny. 1999. Population Dynamics of Northern Yellowstone Mule Deer. In Abstract and Presentation, Northwest Section, The Wildlife Society. March.

Goodman, D. 1996. Viability Analysis of the Antelope Population Wintering Near Gardiner, Montana. Yellowstone Center for Resources, Yellowstone National Park, WY.

Graham, R.W. 1986. Response of mammalian communities to environmental changes during the late Quaternary. Pp. 300-313 in Community Ecology, J. Diamond and T.J. Case, eds. New York: Harper and Row.

Graham, R.W, and E.L. Lundelius, Jr. 1984. Coevolutionary disequilibrium and Pleistocene extinctions. Pp. 223-249 in Quaternary Extinctions: A Prehistoric Revolution, P.S. Martin, and R.G. Klein, eds. Tucson, AZ: University of Arizona Press.

Grant, M.C. 1993. The trembling giant. Discover 14(10):82-89.

Graves, H.S., and E.W. Nelson. 1919. Our National Elk Herds: A Program for Conserving the Elk on National Forests About the Yellowstone National Park. U.S. Dept of Agriculture, Washington, DC.

Grayson, D.K. 1977. On the Holocene history of some northern Great Basin lagomorphs. J. Mammal. 58(4):507-513.

Grayson, D.K. 1981. A mid-Holocene record for the heather vole, Phenacomys cf. intermedius, in the central Great Basin and its biogeographic significance. J. Mammal. 62(1):115-121.

Grayson, D.K. 1982. Toward a history of the Great Basin mammals during the past 15,000 years. Pp. 82-101 in Man and Environment in the Great Basin, D.B. Madsen, and J.F. O'Connell, eds. SAA Papers No. 2. Washington, DC: Society of American Archaeology.

Grayson, D.K. 1987. The biogeographic history of small mammals in the Great Basin: Observations on the last 20,000 years. J. Mammal. 68(2):359-375.

Grayson, D.K. 1993. The Desert's Past: A Natural Prehistory of the Great Basin. Washington, DC: Smithsonian Institution Press. 356 pp.

Greenwood, P.J. 1980. Mating systems, philopatry and dispersal in birds in mammals. Anim. Behav. 28:1140-1162.

Gregory, S.V., F.J. Swanson, W.A. McKee, and K.W. Cummins. 1991. An eco-system perspective of riparian zones. Bioscience 41(8):540-551.

Grimm, R.L. 1938. Northern Yellowstone Range Studies, 1938. Yellowstone National Park Research Library.

Grimm, R.L. 1939. Northern Yellowstone range studies. J. Wildl. Manage 3(4):295-306.

Gross, J.E., L.A. Shipley, N.T. Hobbs, D.E. Spalinger, and B.A. Wunder. 1993. Functional response of herbivores in food-concentrated patches: tests of a mechanistic model. Ecology 74(3):778-791.

Gross, J.E., M.W. Miller, and T.J. Kreeger. 1998. Simulating Dynamics of *Brucellosis* in Elk and Bison. Colorado State University, Colorado Division of Wildlife, and University of Wyoming. Final report to USGS-BRD, Laramie, WY. July 1998. 31 pp.

Gruell, G.E. 1980. Fire's Influence on Wildlife Habitat on the Bridger-Teton National Forest, Wyoming. Vol. 2. Changes and Causes, Management Implications. Research Paper INT-235. Ogden: Department of Agriculture, Forest Service, Intermountain Forest and Range Experiment Station.

Guthrie, D.A. 1971. A primitive man's relationship to nature. Bioscience 21(13):721-723.

Habeck, J.R., and R.W. Mutch. 1973. Fire-dependent forests in the northern Rockies. Quat. Res. 3:408-424.

Hadly-Barnosky, E.A. 1994. Grassland change over 2,000 years in Northern Yellowstone Park. Pp. 319 in Plants and Their Environments: Proceedings of the First Biennial Scientific Conference on the Greater Yellowstone Ecosystem, D. Despain, ed. Tech. Rep. NPS/NRYELL/NRTR. Denver, CO: U.S. Department of the Interior, National Park Service, Natural Resources Publication Office.

Hadly, E.A. 1996. Influence of late Holocene climate on Northern Rocky Mountain mammals. Quat. Res. 46(3):298-310.

Hadly, E.A. 1997. Evolutionary and ecological response of pocket gophers (*Thomomys talpoides*) to late-Holocene climate change. Biol. J. Linnean Soc. 60(2):277-296.

Hagerman, A.E., and C.T. Robbins. 1993. Specificity of tannin-binding salivary proteins relative to diet selection by mammals. Can. J. Zool. 71(3):628-633.

Hanley, T.A., and K.A. Hanley. 1982. Food resource partitioning by sympatric ungulates on Great Basin rangeland. J. Range Manage. 35(2):152-158.

Hansen, A.J., and J.J. Rotella. 1999. Abiotic factors. Pp. 161-209 in Maintaining Biodiversity in Forest Ecosystems, M.L. Hunter Jr, ed. London, UK: Cambridge University Press.

Hansen, A.J., and D.L. Urban. 1992. Avian response to landscape pattern: the role of species' life histories. Landscape Ecol. 7(3)163-180.

Harniss, R.O., and R.B. Murray. 1973. Thirty years of vegetal change following burning sagebrush-grass range. J. Range Manage. 26:322-325.

Harris, A.H. 1978. The Mummy Cave tetrapods. Pp. 146-151 in The Mummy Cave Project in Northwestern Wyoming, H. McCracken, ed. Cody, WY: Buffalo Bill Historical Center.

Harris, R.B., and F.W. Allendorf. 1989. Genetically effective population size of large mammals: an assessment of estimators. Conserv. Biol. 3(2):181-191.

Hayden, J.A. 1989. Status and Population Dynamics of Mountain Goats in the Snake River Range, Idaho. M.S. Thesis, University of Montana, Bozeman. 145 pp.

Hedrick, P.W. 1983. Genetics of Populations. Boston: Science Books International.

Hemming, J.D., and R.L. Lindroth. 1995. Intraspecific variation in aspen phytochemistry: Effects on performance of gypsy moths and forest tent caterpillars. Oecologia 103(1):79-88.

Herriges, J.D., Jr., E.T. Thorne, S.L. Anderson, and H.A. Dawson. 1989. Vaccination of elk in Wyoming with reduced dose strain 19 *Brucella*: controlled studies and ballistic implant field trials. Proc. U.S. Animal Health Assoc. 93:640-655.

Hessl, A.E. 2000. Aspen: Ecological Processes and Management Eras in Northwestern Wyoming, 1807-1998. Ph.D. Dissertation. University of Arizona, Tucson.

Hinds, T.E. 1985. Diseases. Pp. 87-114 in Aspen: Ecology and Management in the Western United States, N.V. DeByle, and R.P. Winokur, eds. Gen. Tech. Rep. RM-119. Fort Collins, CO: U.S. Department of Agriculture Forest Service, Rocky Mountain Forest and Range Experiment Station.

Hinds, T.E., and E.M. Wengert. 1977. Growth and Decay Losses in Colorado Aspen. Forest Service Research Paper RM193. Fort Collins, CO: Rocky Mountain Forest and Range Experiment Station, Forest Service, U.S. Department of Agriculture.

Hobbs, N.T. 1996. Modification of ecosystems by ungulates. J. Wildl. Manage. 60(4):695-713.

Hobbs, N.T., and R.A. Spowart. 1984. Effects of prescribed fire on nutrition of mountain sheep and mule deer during winter and spring. J. Wildl. Manage. 48(2):551-560.

Hoffman, T.L., and C.L. Wambolt. 1996. Growth response of Wyoming big Sagebrush to heavy browsing by wild ungulates. Pp. 242-245 in Proceedings: Shrubland Ecosystem Dynamics in a Changing Environment, Las Cruces, NM, May 23-25, 1995, J.R. Barrow, E.D. McArthur, R.E. Sosebee, and R.E. Rausch, eds. Gen. Tech. Rep. INT-GTR 338. Ogden: Intermountain Research Station, Forest Service, U.S. Department of Agriculture.

Holling, C.S., and G.K. Meffe. 1996. Command and control and the pathology of natural resource management. Conserv. Biol. 10(2):328-337.

Hornaday, W.T. 1889. The extermination of the American bison, with a sketch of its discovery and life history. Pp. 367-552 in United States National Museum, Annual Report of the Board of Regents of the Smithsonian Institution 1887. Washington, DC: U.S. Government Printing Office.

Houston, D.B. 1973. Wildfires in northern Yellowstone National Park. Ecology 54(5):1111-1117.

Houston, D.B. 1982. The Northern Yellowstone Elk: Ecology and Management. New York: Macmillan.

Hudson, R.J., and S. Frank. 1987. Foraging ecology of bison in aspen boreal habitat. J. Range Manage. 40(1):71-75.

Hudson, R.J., D.M. Hebert, and V.C. Brink. 1976. Occupational patterns of wildlife on a major East Kootenay winter-spring range. J. Range Manage. 29(1):38-43.

Huff, D.E., and J.D. Varley. 1999. Natural regulation in Yellowstone National Park's northern range. Ecol. Appl. 9(1):17-29.

Hwang, S.Y., and R.L. Lindroth. 1998. Consequences of clonal variation in aspen phytochemistry for late season folivores. Ecoscience 5(4):508-516.

Jakubas, W.J., G.W. Gullion, and T.P. Clausen. 1989. Ruffed grouse feeding behavior and its relationship to the secondary metabolites of quaking aspen flower buds. J. Chem. Ecol. 15(6):1899-1917.

Jakubos, B., and W.H. Romme. 1993. Invasion of subalpine meadows by lodgepole pine in Yellowstone National Park, Wyoming, USA. Arct. Alp. Res. 25(4):382-390.

Jelinski, D.E., and W.M. Cheliak. 1992. Genetic diversity and spatial subdivision of *Populus tremuloides* (Salicaceae) in a heterogeneous landscape. Am. J. Bot. 79(7):728-736.

Jelinski, D.E., and L.J. Fisher. 1991. Spatial variability in the nutrient composition of *Populus tremuloides*: clone-to-clone differences and implications for cervids. Oecologia 88(1):116-124.

Johnson, E.A. 1992. Fire and Vegetation Dynamics: Studies from the North American Boreal Forest. Cambridge: Cambridge University Press.

Johnson, E.A., and G.I. Fryer. 1987. Historical vegetation change in the Kananaskis Valley, Canadian Rockies. Can. J. Bot. 65(5):853-858.

Johnson, K.A., and R.L. Crabtree. 1999. Small prey of carnivores in the Greater Yellowstone Ecosystem. Pp. 239-264 in Carnivores in Ecosystems: The Yellowstone Experience, T.W. Clark, A.P. Curlee, S.C. Minta, and P.M. Kareiva, eds. New Haven, CT: Yale University Press.

Jonas, R.J. 1955. A Population and Ecological Study of the Beaver (Castor Canadensis) in Yellowstone National Park. M.S. Thesis. University of Idaho. 193 pp.

Jones, J.R., and N.V. DeByle. 1985. Genetics and variation. Pp. 35-39 in Aspen: Ecology and Management in the Western United States, N.V. DeByle, and R.P. Winokur, eds. Gen. Tech. Rep. RM-119. Fort Collins, CO: U.S. Department of Agriculture Forest Service, Rocky Mountain Forest and Range Experiment Station.

Jonkel, C.J. 1987. Brown bear. Pp. 456-473 in Wild Furbearer Management and Conservation in North America, M. Novak, J.A. Baker, M.E. Obbard, and B. Malloch, eds. Ontario: Ministry of Natural Resources.

Kaufmann, J.B., D. Cummings, C. Heider, D. Lytjen, and N. Otting. 2000. Riparian vegetation responses to re-watering and cessation of grazing, Mono Basin, California. Pp. 251-256 in Riparian Ecology and Management in Multi-Land Use Watersheds. Proceedings AWRA 2000 Summer Special Conference, August 28-31, Portland, OR, P.J. Wigington, Jr., and R.L. Beschta, eds. Middleburg, VA: American Water Resources Association.

Kay, C.E. 1990. Yellowstone's Northern Elk Herd: A Critical Evaluation of the "Natural-Regulation" Paradigm. Ph.D. dissertation. Utah State University, Logan.

Kay, C.E. 1993. Aspen seedlings in recently burned areas of Grand Teton and Yellowstone National Parks. Northwest Sci. 67(2):94-104.

Kay, C.E. 1994. The impact of native ungulates and beaver on riparian communities in the Intermountain West. Natural Resources and Environmental Issues 1:23-44.

Kay, C.E. 1995. Aboriginal overkill and native burning: implications for modern ecosystem management. Pp. 107-118 in Sustainable Society and Protected Areas, Contributed Papers of the 8th Conference on Research and Resources Management in Parks and on Public Lands. April 17-21, 1995, Portland, OR, R.M. Linn, ed. Hancock, MI: The George Wright Society.

Kay, C.E. 1997. The condition and trend of aspen, Populus tremuloides, in Kootenay and Yoho National Parks: implications for ecological integrity. Can. Field Natur. 111(4):607-616.

Kay, C.E., and S.W. Chadde. 1992. Reduction in willow seed production by ungulate browsing in Yellowstone National Park. Pp. 92-99 in Proceedings- Symposium of Ecology and Management of Riparian Shrub Communities, Sun Valley, ID, May 29-31, 1991, W.P. Clary, E.D. McArthur, D. Bedunah, and C.L. Wambolt, eds. Ogden, UT: Intermountain Research Station, Forest Service, U.S. Department of Agriculture.

Kay, C.E., and F.H. Wagner. 1994. Historical condition of woody vegetation on Yellowstone's Northern Range: a critical evaluation of the "Natural Regulation" Paradigm. Pp. 151-169 in Plants and Their Environments: Proceedings of the First Biennial Scientific Conference on the Greater Yellowstone Ecosystem, D. Despain, ed. Tech. Rep. NPS/NRYELL/NRTR. Denver, CO: U.S. Department of the Interior, National Park Service, Natural Resources Publication Office.

Kay, C.E., and F.H. Wagner. 1996. Response of shrub-aspen to Yellowstone's 1988 wildfires: implications for "Natural Regulation" management. Pp. 107-111 in The Ecological Implications of Fire in Greater Yellowstone, Proceedings Second Biennial Conference on the Greater Yellowstone Ecosystem, Yellowstone National Park, Sept. 19-21, 1993, J.M. Greenlee, ed. Fairfield, WA: International Association of Wildland Fire.

Keigley, R.B. 1997a. A growth form method for describing browse condition. Rangelands 19(3):26-29.

Keigley, R.B. 1997b. An increase in herbivory of cottonwood in Yellowstone National Park. Northwest Sci. 71(2):127-136.

Keigley, R.B. 1998. Architecture of cottonwood as an index of browsing history in Yellowstone. Interm. J. Sci. 4(3/4):57-67.

Keigley, R.B. 2000. Elk, beaver, and the persistence of willows in national parks: comment on Singer et al. (1998). Wildl. Soc. Bull. 28(2):448-450.

Keigley, R.B., and F.H. Wagner. 1998. What is "natural"? Yellowstone elk population —A case study. Integr. Biol. 1:134-148.

Kendall, K.C. 1983. Use of pine nuts by grizzly and black bears in the Yellowstone area. Pp. 166-173 in Bears-Their Biology and Management. Papers of the Fifth International Conference on Bear Research and Management, Feb. 1980, Madison, WI., E.C. Meslow, ed. Madison, WI: International Association for Bear Research and Management.

Kendall, K.C. 1998. Whitebark pine. Pp. 483-485 in Status and Trends of the Nations Biological Resources, Vol. 2. Regional Trends of Biological Resources, M.J. Mack, P.A. Opler, C.E. Pockett, and P.D. Doran, eds. Reston, VA: U.S. Geological Survey. [Online]. Available: http://biology.usgs.gov/s+t/SNT/noframe/wm147.htm [October 5, 2001].

Kendall, K.C., and J.M. Asebrook. 1998. The war against Blister Rust in Yellowstone National Park, 1945-1978. The George Wright Forum 15(4):36-49.

Kendall, K.C., and R.E. Keane. 2001. Whitebark pine decline: infection, mortality, and population trends. Pp. 221-242 in Whitebark Pine Communities: Ecology and Restoration, D.F. Tomback, S.F. Arno, and R.E. Keane, eds. Washington, DC: Island Press.

Kendall, K.C., and D. Schirokauer. 1997. Alien threats and restoration dilemmas in white and limber pine communities. Pp. 218-225 in Making Protection Work: Proceedings of the 9th Conference on Research and Resources Management in Parks and Public Lands; the George Wright Foundation Biennial Conference, Albuquerque, NM, March 17-21, 1997. Hancock, MI: The George Wright Society.

Kittams, W.H. 1952. Northern Winter Range Studies: Yellowstone National Park, Vols. 2 and 3. Washington, DC: U.S. Department. of the Interior, National Park Service.

Klein, D.R. 1995. The introduction, increase, and demise of wolves on Coronation Island, Alaska. Pp. 275-280 in Ecology and Conservation of Wolves in a Changing World, L.N. Carbyn, S.H. Fritts, and D.R. Seip, eds. Edmonton, Alberta: Canadian Circumpolar Institute.

Klein, D.R., D.F. Murray, R.H. Armstrong, and B.A. Anderson. 1998. Alaska. Pp. 708-745 in Status and Trends of the Nation's Biological Resources, Vol. 2. Regional Trends of Biological Resources, M.J. Mac, P.A. Opler, C.E. Puckett, and P.D. Doran, eds. Reston, VA: U.S. Geological Survey. [Online]. Available: http://biology.usgs.gov/s+t/SNT/index.htm [September 9, 2001].

Knight, D.H. 1987. Parasites, lightning, and the vegetation mosaic in wilderness landscapes. Pp. 59-83 in Landscape Heterogeneity and Disturbance, M.G. Turner, ed. New York: Springer-Verlag.

Knight, D.H. 1994. Mountains and Plains: The Ecology of Wyoming Landscapes. New Haven: Yale University Press.

Knight, D.H. 1996. The ecological implications of fire in Greater Yellowstone: A summary. Pp. 233-235 in Ecological Implications of Fire in Greater Yelllowstone: Proceedings Second Biennial Conference on the Greater Yellowstone Ecosystem, Yellowstone National Park, September 19-21, 1993, J.M. Greenlee, ed. Fairfield, WA: International Association of Wildland Fire.

Knight, R.R., and L.L. Eberhardt. 1985. Population dynamics of Yellowstone grizzly bears. Ecology 66(2):323-334.

Knight, R.R., B.M. Blanchard, and P. Schullery. 1999. Yellowstone bears. Pp. 51-76 in Carnivores in Ecosystems: The Yellowstone Experience, T.W. Clark, A.P. Curlee, S.C. Minta, and P.M. Kareiva, eds. New Haven, CT: Yale University Press.

Kolenosky, G.B., and S.M. Strathearn. 1987. Black bear. Pp. 442-455 in Wild Furbearer Management and Conservation in North America, M. Novak, J.A. Baker, M.E. Obbard, and B. Malloch, eds. Ontario: Ministry of Natural Resources.

Krebill, R.G. 1972. Mortality of Aspen on the Gros Ventre Elk Winter Range. Forest Service Research Paper INT-129. Ogden, UT: Intermountain Forest and Range Experiment Station, Forest Service, U.S. Department of Agriculture.

Krech, S. 1999. The Ecological Indian: Myth and History. New York: W.W. Norton. 318 pp.

Kurten, B., and E. Anderson. 1980. Pleistocene Mammals of North America. New York: Columbia University Press. 442 pp.

Laird, K.R., S.C. Fritz, E.C. Grimm, and P.G. Mueller. 1996. Century-scale paleoclimatic reconstruction from Moon Lake, a closed-basin lake in the northern Great Plains. Limnol. Oceanogr. 41(5):890-902.

Landres, P.B., P. Morgan, and F.J. Swanson. 1999. Overview of the use of natural variability concepts in managing ecological systems. Ecol. Appl. 9(4):1177-1188.

Lane, J.R., and C. Montagne. 1996. Characterization of soils from grazing exclosures and adjacent areas in northern Yellowstone National Park. Pp. 63-72 in Effects of Grazing by Wild Ungulates in Yellowstone National Park, F.J. Singer, ed. Tech. Rep. NPS/NRYELL/NRTR/96-01. U.S. Department. of the Interior, National Park Service, Natural Resource Information Division, Denver, CO.

Laundre, J.W. 1990. The Status, Distribution, and Management of Mountain Goats in The Greater Yellowstone Ecosystem. Final report to Yellowstone National Park, NPS Order PX 1200-8-0828, Department of Biological Science, Idaho State University, Pocatello. 58 pp.

Laundre, J.W. 1994. Resource overlap between mountain goats and bighorn sheep. Great Basin Nat. 54(2):114-121.

Lee, T.E., Jr., J.W. Bickham, and M.D. Scott. 1994. Mitochondrial DNA and allozyme analysis of North American pronghorn populations. J. Wildl. Manage. 58(2):307-318.

Lemke, T. 1998. Gardiner Late Elk Hunt Annual Report. Montana Department of Fish, Wildlife and Parks. Livingston, MT. 62 pp.

Lemke, T. 1999. Northern Yellowstone Cooperative Wildlife Working Group, 1999 Annual Report. National Park Service, Yellowstone National Park; Montana Department of Fish, Wildlife and Parks, Gallatin National Forest, U.S. Forest Service, and Biological Resources Division, U.S. Geological Survey.

Lemke, T.O., J.A. Mack, and D.B. Houston. 1998. Winter range expansion by the northern Yellowstone elk herd. Interm. J. Sci. 4(1/2):1-9.

Leopold, A. 1933. Game Management. New York: Charles Scribner's Sons. 481 pp.

Leopold, A.S., S.A. Cain, C.M. Cottam, I.N. Gabrielson, and T.L. Kimball. 1963. Study of wildlife problems in the national parks. Trans. North Amer. Wildl. Nat. Resour. Conf. 28:28-45.

Leopold, L.B., and M.G. Wolman. 1957. River Channel Patterns: Braided, Meandering, and Straight. U.S. Geological Survey Professional Paper 282-B. Washington, DC: U.S. Government Printing Office.

Lindroth, R.L., and S. Hwang. 1996. Clonal variation in foliar chemistry of quaking aspen (*Populus tremuloides* Michx.). Biochem. Syst. Ecol. 24(5):357-364.

Lindzey, F. 1987. Mountain lion. Pp. 656-669 in Wild Furbearer Management and Conservation in North America, M. Novak, J.A. Baker, M.E. Obbard, and B. Malloch, eds. Ontario: Ministry of Natural Resources.

Loope, L.L., and G.E. Gruell. 1973. The ecological role of fire in the Jackson Hole area, northwestern Wyoming. Quat. Res. 3:425-443.

Lyons, J., S.W. Trimble, and L.K. Paine. 2000. Grass versus trees: managing riparian areas to benefit streams of central North America. J. Am. Water Resour. Assoc. 36(4):919-930.

Mack, J.A., and F.A. Singer. 1993. Using POP-II models to predict effects of wolf predation and hunter harvests on elk, mule deer, and moose on the Northern Range. Pp. 49-74 in Ecological Issues on Reintroducing Wolves into Yellowstone National Park, R.A. Cook, ed. National Parks Service Scientific Monograph NPS/NRYELL/NRSM-93/22. Denver, CO: National Park Services, U.S. Department of the Interior.

Magoc, C.J. 1999. Yellowstone: The Creation and Selling of an American Landscape, 1870-1903. Albuquerque: University of New Mexico Press. 266 pp.

Malanson, G.P. 1993. Riparian Landscapes. Cambridge: Cambridge University Press.

Martin, L.D., and B.M. Gilbert. 1978. Excavations at Natural Trap Cave. Trans. Nebraska Acad. Sci. 6:107-116.

Martin, P., and R. Klein, eds. 1984. Quaternary Extinctions: A Prehistoric Revolution. Tucson, AZ: University of Arizona Press.

Martinka, C.J. 1967. Mortality of northern Montana pronghorns in a severe winter. J. Wildl. Manage. 31(1):159-164.

Mattson, D.J., B.M. Blanchard,.and R.R. Knight. 1991. Food habits of Yellowstone grizzly bears, 1977-87. Can. J. Zool. 69(6):1619-1629.

Mattson, D.J., K.C. Kendall, and D.P. Reinhart. 2001. Whitebark pine, grizzly bears and red squirrels. Pp. 121-136 in Whitebark Pine Communities: Ecology and Restoration, D.F. Tomback, S.F. Arno, and R.E. Keane, eds. Washington, DC: Island Press.

McCullough, D.R. 1979. The George Reserve Deer Herd: Population Ecology of a K-Selected Species. Ann Arbor: University of Michigan Press.

McCollough, D.R. 1990. Detecting density dependence: filtering the baby from the bathwater. Trans. 55th N. Amer. Wildl. Nat. Resour. Conf. 55:534-543.

McCullough, D.R. 1992. Concepts of large herbivore population dynamics. Pp. 967-984 in Wildlife 2001: Populations, D.R. McCullough, and R.H. Barrett, eds. New York: Elsevier Applied Science.

McCullough, Y.B. 1980. Niche Separation of Seven North American Ungulates on the National Bison Range, Montana. Ph.D. Thesis. University of Michigan.

McDonough, W.T. 1985. Sexual reproduction, seeds and seedlings. Pp. 25-33 in Aspen: Ecology and Management in the Western United States, N.V. DeByle, and R.P. Winokur, eds. Gen. Tech. Rep. RM-119. Fort Collins, CO: U.S. Department of Agriculture Forest Service, Rocky Mountain Forest and Range Experiment Station.

McLaren, B.E., and R.O. Peterson. 1994. Wolves, moose, and tree rings on Isle Royale. Science 266(5190):1555-1558.

Mead, J.I. 1987. Quaternary records of pika, Ochotona, in North America. Boreas 16:165-171.

Mead, J.I., R.S. Thompson, and T.R. Van Devender. 1982. Late Wisconsinan and Holocene fauna from Smith Creek Canyon, Snake Range, Nevada. Trans. San Diego Soc. Nat. Hist. 20:1-26.

Meagher, M.M. 1973. The Bison of Yellowstone National Park. National Park Service Science Monograph Series 1. Washington DC. 161 pp.

Meagher, M.M. 1989. Range expansion by bison of Yellowstone National Park. J. Mammal. 70(3):670-675.

Meagher, M.M., and D.B. Houston. 1998. Yellowstone and the Biology of Time: Photographs Across a Century. Norman: University of Oklahoma Press.

Meagher, M., and M.E. Meyer. 1994. On the origin of brucellosis in bison of Yellowstone National Park: A review. Conserv. Biol. 8(3):645-653.

Mech, L.D., T.J. Meier, J.W. Burch, and L.G. Adams. 1995. Patterns of prey selection by wolves in Denali National Park, Alaska. Pp. 231-243 in Ecology and Conservation of Wolves in a Changing World, L.N. Carbyn, S.H. Fritts, and D.R. Seip, eds. Edmonton, Alberta: Canadian Circumpolar Institute.

Merriam-Webster. 1993. Merriam-Webster Collegiate Dictionary, 10th Ed. Springfield, MA: Merriam-Webster, Inc.

Merrill, E.H., and M.S. Boyce. 1991. Summer range and elk population dynamics in Yellowstone National Park. Pp. 263-274 in The Greater Yellowstone Ecosystem: Redefining America's Wilderness Heritage, R.B. Keiter and M.S. Boyce, eds. New Haven: Yale University Press.

Messier, F. 1994. Ungulate population models with predation: a case study with the North American moose. Ecology 75(2):478-488.

Messier, F. 1995. On the functional and numerical responses of wolves to changing prey density. Pp. 187-197 in Ecology and Conservation of Wolves in a Changing World, L.N. Carbyn, S.H. Fritts, and D.R. Seip, eds. Edmonton, Alberta: Canadian Circumpolar Institute.

Meyer, G.A., S.G. Wells, R.C. Balling, Jr., and A.J.T. Jull. 1992. Response of alluvial systems to fire and climate change in Yellowstone National Park. Nature 357(6374): 147-149.

Meyer, G.A., S.G. Wells, and A.J.T. Jull. 1995. Fire and alluvial chronology in Yellowstone National Park: climatic and intrinsic controls on Holocene geomorphic processes. Geol. Soc. Am. Bull. 107(10):1211-1230.

Miller, K.G., R.G. Fairbanks, and G.S. Mountain. 1987. Tertiary oxygen isotope synthesis, sea level history, and continental margin erosion. Paleoceanography 20(1):1-19.

Miller, S.D., G.C. White, R.A. Sellers, H.V. Reynolds, J.W. Schoen, K. Titus, V.G. Barnes, Jr., R.B. Smith, R.R. Nelson, W.B. Ballard, and C.C. Schwartz. 1997. Brown and Black Bear Density Estimation in Alaska Using Radiotelemetry and Replicated Mark-Resight Techniques. Wildlife Monographs No. 133. 55 pp.

Millspaugh, S.H., and C. Whitlock. 1995. A 750-year fire history based on lake sediment records in central Yellowstone National Park, USA. The Holocene 5(3):283-292.

Miquelle, D.G., J.M. Peek, and V. van Balenberghe. 1992. Sexual Segregation in Alaskan Moose. Wildlife Monographs No. 122. Bethesda, MD: Wildlife Society. 57 pp.

Mohler, J.R. 1917. Annual Report of the U.S. Bureau of Animal Industry. Washington, DC: U.S. Department of Agriculture. 106 pp.

Montana Department of Livestock and Montana Fish, Wildlife and Parks. 2000. Interagency Bison Management Plan for The State of Montana and Yellowstone National Park. Final Environmental Impact Statement. Montana Department of Livestock and Montana Fish, Wildlife and Parks. November 15, 2000. [Online]. Available: http://www.liv.state.mt.us/BISON/MANPLAN11-15/ [October 17, 2001].

Morton, J.K., E.T. Thorne, and G.M. Thomas. 1981. Brucellosis in elk. III. Serologic evaluation. J. Wildl. Dis. 17(1):23-31.

Moss, E.H. 1938. Longevity of seed and establishment of seedlings in species of Populus. Bot. Gaz. 99(3):529-542.

Mowry, A.D. 1998. Vegetation and Channel Changes on Soda Butte Creek, Yellowstone National Park. Geology Senior Theses. Middlebury College, Middlebury, VT.

Mueggler, E.H., and W.L. Stewart. 1980. Grassland and Shrubland Habitat Types of Western Montana. Gen. Tech. Rep. INT 66. Ogden, UT: U.S. Department of Agriculture, Forest Service, Intermountain Research Station. 154 pp.

Mueggler, W.F. 1988. Aspen Community Types of the Intermountain Region. Gen. Tech. Rep. INT 250. Ogden, UT: U.S. Department of Agriculture, Forest Service, Intermountain Research Station.

Muller, R.A., and G.J. MacDonald. 1997. Glacial cycles and astronomical forcing. Science 277(5323):215-218.

Murie, A. 1940. Fauna of the National Parks of the United States. Ecology of the Coyote in the Yellowstone. Washington, DC: U.S. Government Printing Office.

Murie, O.J. 1951. The Elk of North America, 1st Ed. Harrisburg, PA: Stackpole. 376 pp.

Murphy, K.M. 1998. The Ecology of the Cougar (*Puma concolor*) in the Northern Yellowstone Ecosystem: Interactions with Prey, Bears, and Humans. Ph.D. Dissertation. University of Idaho, Moscow.

Murphy, K.M., P.I. Ross, and M.G. Hornocker. 1999. The ecology of anthropogenic influences on cougars. Pp. 77-102 in Carnivores in Ecosystems: The Yellowstone Experience, T.W. Clark, A.P. Curlee, S.C. Minta, and P.M. Kareiva, eds. New Haven, CT: Yale University Press.

Naiman, R.J., and H. Decamps. 1997. The ecology of interfaces: riparian zones. Annu. Rev. Ecol. Syst. 28(1):621-658.

Naiman, R.J., H. Decamps, and M. Pollock. 1993. The role of riparian corridors in maintaining regional biodiversity. Ecol. Appl. 3(2):209-212.

Noss, R.F., and A.Y. Cooperrider. 1994. Saving Nature's Legacy: Protecting and Restoring Biodiversity. Washington, DC: Island Press. 416 pp.

NPS (National Park Service). 1998. Draft Environmental Impact Statement for the Interagency Bison Management Plan for the State of Montana and Yellowstone National Park. Denver, CO: National Park Service.

NPS (National Park Service). 2001. 2001 Management Policies. [Online]. Available: http://www.nps.gov/refdesk/mp/ [Nov. 16, 2001].

NRC (National Research Council). 1997. Wolves, Bears, and Their Prey in Alaska. Washington, DC: National Academy Press.

NRC (National Research Council). 1998. Brucellosis in the Greater Yellowstone Area. Washington, DC: National Academy Press.

NRC (National Research Council). 2001. Climate Change Science, An Analysis of Some Key Questions. Washington, DC: National Academy Press.

O'Gara, B.W. 1968. A Study of the Reproductive Cycle of the Female Pronghorn (Antilocapra americana Ord). Ph.D. Dissertation, University of Montana, Missoula.

Olmstead, C.E. 1979. The ecology of aspen with reference to utilization by large herbivores in Rocky Mountain National Park. Pp. 89-97 in North American Elk: Ecology, Behavior, and Management, M.S. Boyce, and L.D. Hayden-Wing, eds. University of Wyoming, Laramie.

Paetkau, D., L.P. Waits, P.L. Clarkson, L. Craighead, E. Vyse, R. Ward, and C. Strobeck. 1998. Variation in genetic diversity across the Range of North American brown bears. Conserv. Biol. 12(2)418-429.

Paine, R.T. 1966. Food web complexity and species diversity. Am. Nat. 100(910):65-75.

Palo, R.T. 1984. Distribution of birch (*Betula* spp.), willow (*Salix* spp.), and poplar (*Populus* spp.) secondary metabolites and their potential role as chemical defense against herbivores. J. Chem. Ecol. 10(3):499-520.

Park County. 2001. Park County Montana. [Updated Nov. 19, 2001]. [Online]. Available: www.parkcounty.org [February 22, 2002].

Patten, D.T. 1963. Vegetation patterns in relation to environments in the Madison Range, Montana. Ecol. Monogr. 33(4):375-406.

Patten, D.T. 1968. Dynamics of the shrub continuum along the Gallatin River in Yellowstone National Park. Ecology 49(6):1107-1112.

Patten, D.T. 1969. Succession from sagebrush to mixed conifer forest in the northern Rocky Mountains. Am. Midland Nat. 82(1):229-240.

Patten, D.T. 1991. Human impacts in the Greater Yellowstone ecosystem: evaluating sustainability goals and eco-redevelopment. Conserv. Biol. 5(3):405-411.

Patten, D.T. 1998. Riparian ecosystems of semi-arid North America: diversity and human impacts. Wetlands 18(4):498-512.

Patten, D.T. 2000. Riparian ecosystems of North America's intermountain west and adjacent plains. Pp. 245-250 in Riparian Ecology and Management in Multi-Land Use Watersheds: Proceedings, AWRA's 2000 Summer Specialty Conference, August 28-31, 2000, Portland, OR, P.J. Wigington, and R.L. Beschta, eds. TPS00-2. Middleburg, VA: American Water Resources Association.

Peacock, D. 1997. The Yellowstone massacre. Audubon 93(3):40-49, 102-103, 106-107.

Pearson, S.M., M.G. Turner, L.L. Wallace, and W.H. Romme. 1995. Winter habitat use by large ungulates following fires in northern Yellowstone National Park. Ecol. Appl. 5(3):744-755.

Pease, C.M., and D.J. Mattson. 1999. Demography of the Yellowstone grizzly bears. Ecology 80(3):957-975.

Pengelly, W.L. 1961. Factors Influencing Production of White-Tailed Deer on the Coeur d'Alene National Forest, Idaho. Ph.D. Dissertation. Utah State University, Logan.

Pepper, G.W., and R. Quinn. 1965. Antelope Population Trend Survey in Saskatchewan 1965. Saskatchewan Department of Natural Resources Report. 11 pp.

Peterson, M.J., W.E. Grant, and D.S. Davis. 1991. Simulation of host-parasitic interactions within a resource management framework: impact of brucellosis on bison population dynamics. Ecol. Model. 54(3/4):299-320.

Petit, J.R., J. Jouzel, D. Raynaud, N.I. Barkov, J.M. Barnola, I. Basile, M. Bender, H. Chappellaz, M. Davis, G. Delayque, M. Delmotte, V.M. Kotlyakov, M. Legrand, V.Y. Lipenkov, C. Lorius, L. Pepin, C. Ritz, E. Saltzman, and M. Stievenard. 1999. Climate and atmospheric history of the past 420,000 years from the Vostok ice core, Antarctica. Nature 399(735):429-436.

Pfitsch, W., R.S. Reid, J. Harter, D.K. Pike, and L.C. Bliss. 1983. Effects of Mountain Goats on Soils, Plant Communities, and Select Species in Olympic National Park. Final Report. Seattle, WA: Department of Botany and College of Forest Resources, University of Washington.

Pierce, K.L., and L.A. Morgan. 1992. The track of the Yellowstone hot spot: volcanism, faulting, and uplift. Geol. Soc. Amer. Memoir 179:1-53.

Pojar, R.A. 1995. Breeding Bird Communities in Aspen Forests of the Sub-Boreal Spruce (dk subzone) in the Prince Rupert Forest Region. Land Management Handbook 33. Victoria, B.C: Province of British Columbia, Ministry of Forests Research Program.

Pollard, E., K.H. Lakhani, and P. Rothery. 1987. The detection of density-dependence from a series of annual censuses. Ecology 68(6):2046-2055.

Post, E., R.O. Peterson, N.C. Stenseth, and B.E. McLaren. 1999. Ecosystem consequences of wolf behavioural response to climate. Nature 401(6756):905-907.

Pregitzer, K.S., and A.L. Friend. 1996. The structure and function of *Populus* root systems. Pp. 331-354 in Biology of Populus and Its Implications for Management and Conservation, R.F. Stettler, H.D. Bradshaw, Jr., P.E. Heilman, and T.M. Hinckley, eds. Ottawa: NRC Research Press, National Research Council of Canada.

Pritchard, J.A. 1999. Preserving Yellowstone's Natural Conditions. Lincoln, NE: University of Nebraska Press. 370 pp.

Purdue, J.R. 1989. Changes during the Holocene in the size of white-tailed deer (Odocoileus virginianus) from Central Illinois. Quat. Res. 32(3)307-316.

Reichardt, P.B., J.P. Bryant, B.R. Mattes, T.P. Clausen, F.S. Chapin, and M. Meyer. 1990. Winter chemical defense of Alaskan balsam poplar against snowshoe hares. J. Chem. Ecol. 16(6):1941-1959.

Reilinger, R.E. 1985. Vertical movements associated with the 1959, M=7.1 Hebgen Lake, Montana earthquake. Pp. 519-530 in Proceedings of Workshop XXVIII on the Borah Peak, Idaho, Earthquake, Vol. A, R.S. Stein, and R.C. Bucknam, eds. Menlo Park, CA: U.S. Geological Survey.

Renecker, L.A., and R.J. Hudson. 1988. Seasonal quality of forages used by moose in the aspen-dominated boreal forest, central Alberta. Holartic Ecol. 11(2):111-118.

Renkin, R.A., and D.G. Despain. 1992. Fuel moisture, forest type and lightning-caused fire in Yellowstone National Park. Can. J. For. Res. 22(1):37-45.

Renkin, R., and D. Despain. 1996. Notes on postfire aspen seedling establishment. Pp. 105-106 in Ecological Implications of Fire in Greater Yellowstone, Proceedings Second Biennial Conference on the Greater Yellowstone Ecosystem, Yellowstone National Park, Sept. 19-21, 1993, J.M. Greenlee, ed. Fairfield, WA: International Association of Wildland Fire.

Reynolds, H.W., R.D. Glaholt, and A.W.L. Hawley. 1982. Bison (*Bison bison*). Pp. 972-1007 in Wild Mammals of North America: Biology, Management, and Economics, J.A. Chapman, and G.A. Feldhamer, eds. Baltimore: Johns Hopkins University Press.

Rhyan, J.C., S.D. Holland, T. Gidlewski, D.A. Saari, A.E. Jensen, D.R. Ewalt, S.G. Hennager,

S.C. Olsen, and N.F. Cheville. 1997. Seminal vesiculitis and orchitis caused by Brucella abortus biovar 1 in young bison bulls from South Dakota. J. Vet. Diagn. Invest. 9(4): 368-374.

Ripple, W.J., and E.J. Larsen. 2000b. Historic aspen recruitment, elk, and wolves in northern Yellowstone National Park. Biol. Conserv. 95(3):361-370.

Ripple, W.J., and E.J. Larsen. 2000a. Remote sensing of aspen change in northern Yellowstone National Park. Pp. 24-25 in 1997-1998 Investigators' Annual Reports, Yellowstone National Park, D. Schneider and A. Deutch, eds. YCR-IAR-1997-98. Yellowstone Center for Resources, Yellowstone National Park, WY. April.

Ripple, W.J., and E.J. Larsen. 2001. The role of postfire coarse woody debris in aspen regeneration. Western J. Appl. For. 16(2):61-64.

Robbins, R.L., D.E. Redfearn, and C.P. Stone. 1982. Refuges and elk management. Pp. 479-507 in Elk of North America: Ecology and Management, J.W. Thomas, and D.E. Toweill, eds. Harrisburg, PA: Stackpole Books.

Romme, W.H. 1982. Fire and landscape diversity in subalpine forests of Yellowstone National Park. Ecol. Monogr. 52(2):199-221.

Romme, W.H., and D.G. Despain. 1989. Historical perspective on the Yellowstone fires of 1988. BioScience 39(2):695-699.

Romme, W.H., and D.H. Knight. 1981. Fire frequency and subalpine forest succession along a topographic gradient in Wyoming. Ecology 62(2):319-326.

Romme, W.H., and D.H. Knight. 1982. Landscape diversity: the concept applied to Yellowstone Park. BioScience 32(2):664-670.

Romme, W.H., and M.G. Turner. 1991. Implications of global climate change for biogeographic patterns in the Greater Yellowstone ecosystem. Conserv. Biol. 5(3):373-386.

Romme, W.H., L. Floyd-Hanna, D.D. Hanna, and E. Bartlett. 2001. Aspen's ecological role in the west. Pp. 243-259 in Sustaining Aspen in Western Landscapes: Symposium Proceedings, June 13-15, 2000, Grand Junction, Colorado. Proceedings RMRS-P-18. Fort Collins, CO: U.S. Department of Agriculture, Forest Service, Rocky Mountain Research Station. [Online]. Available: http://www.fs. fed.us/rm/pubs/rmrs_p18.html [Nov. 16, 2001].

Romme, W.H., M.G. Turner, R.H. Gardner, W.W. Hargrove, G.A. Tuskan, D.G. Despain and R. Renkin. 1997. A rare episode of sexual reproduction in aspen (*Populus tremuloides* Michx.) following the 1988 Yellowstone fires. Natural Areas Journal 17(1):17-25.

Romme, W.H., M.G. Turner, L.L. Wallace, and J. Walker. 1995. Aspen, elk, and fire in northern Yellowstone National Park. Ecology 76(7):2097-2106.

Rood, S.B., and J.M Mahoney. 1995. River damming and riparian cottonwoods along the Marias River, Montana. Rivers 5(3):195-207.

Rosgen, D.L. 1993. Stream Classification, Streambank Erosion and Fluvial Interpretations for the Lamar River and Main Tributaries. Final contract report to the National Park Service. Wildland Hydrology, Pagosa Springs, CO. 22 pp.

Ross, P.I., M.G. Jalkotzy, and M. Festa-Bianchet. 1997. Cougar predation on bighorn sheep in southwest Alberta during winter. Can. J. Zool. 75(5):771-775.

Rothstein, S.I. 1994. The cowbird's invasion of the Far West: history, causes, and consequences experienced by host species. Pp. 301-315 in A Century of Avifaunal Change in Western North America, J.R. Jehl, and N.K. Johnson, eds. Camarillo, CA: Cooper Ornithological Society.

Royama, T. 1992. Analytical Population Dynamics, 1st Ed. New York: Chapman and Hall. Schier, G.A. 1975. Deterioration of Aspen Clones in the Middle Rocky Mountains.

Forest Service Research Paper INT-170. Ogden, Utah: Intermountain Forest and Range Experiment Station, Forest Service, U.S. Department of Agriculture.

Schullery, P. 1992. The Bears of Yellowstone. Worland, WY: High Plains Pub.

Schullery, P. 1997. Searching for Yellowstone: Ecology and Wonder in the Last Wilderness. Boston: Houghton Mifflin.

Schullery, P., and L.H. Whittlesey. 1992. Documentary record of wolves and related wildlife species in the Yellowstone National Park area prior to 1882. Pp. 1-173 in Wolves for Yellowstone? A Report to the United States Congress, Vol. 4. Research and Analysis, J.D. Varley, and W.G. Brewster, eds. National Park Service, Yellowstone National Park. July.

Schullery, P., and L.H. Whittlesey. 1999. Greater Yellowstone carnivores: a history of changing attitudes. Pp. 11-49 in Carnivores in Ecosystems: The Yellowstone Experience, T.W. Clark, A.P. Curlee, S.C. Minta, and P.M. Kareiva, eds. New Haven, CT: Yale University Press.

Schwartz, C.C., and J.E. Ellis. 1981. Feeding ecology and niche separation in some native and domestic ungulates on the shortgrass prairie. J. Appl. Ecol. 18(2):343-353.

Schwartz, C.C., J.G. Nagy, and R.W. Rice. 1977. Pronghorn dietary quality relative to forage availability and other ruminants in Colorado. J. Wildl. Manage. 41(2):161-168.

Scott, D.M. 1991. Pronghorn Study continues near YNP's north boundary. The Buffalo Chip, Resource Management Newsletter, Yellowstone National Park. September-October.

Scott, D.M. 1992. Buck-and-pole fence crossings by 4 ungulate species. Wildl. Soc. Bull. 20(2):204-210.

Scott, M.L., G.T. Auble, and J.M. Friedman. 1997. Flood dependency of cottonwood establishment along the Missouri River, Montana. Ecol. Appl. 7(2):677-690.

Sellars, R.W. 1997. Preserving Nature in the National Parks: A History. New Haven: Yale University Press.

Semken, H.A., Jr., and R.W. Graham. 1996. Paleoecologic and taphonomic patterns derived from correspondence analysis of zooarcheological and paleontological faunal samples, a case study from the North American prairie/forest ecotone. Acta Zoologica Cracoviensia 39(1):477-490.

Senft, R.L., M.B. Coughenour, D.W. Bailey, L.R. Rittenhouse, O.E. Sala, and D.M. Swift. 1987. Large herbivore foraging and ecological hierarchies. BioScience 37(11):789-795, 798-799.

Shenk, T.M., G.C. White, and K.P. Burnham. 1998. Sampling variance effects on detecting density dependence from temporal trends in natural populations. Ecol. Monogr. 68(3):445-463.

Shipley, L.A., S. Blomquist, and K. Danell. 1998. Diet choices made by free-ranging moose in northern Sweden in relation to plant distribution, chemistry, and morphology. Can. J. Zool. 76(9):1722-1733.

Sinclair, A.R.E. 1979. Dynamics of the Serengati ecosystem: process and pattern. Pp. 1-30 in Serengeti: Dynamics of an Ecosystem, A.R.E. Sinclair, and M. Norton-Griffiths, eds. Chicago, IL: The University of Chicago Press.

Sinclair, A.R.E. 1989. The regulation of animal populations. Pp. 197-241 in Ecological Concepts, J.M. Cherrett, ed. Oxford: Blackwell.

Sinclair, A.R.E., and P. Arcese. 1995. Population consequences of predation-sensitive foraging: The Serengeti wildebeest. Ecology 76(3):882-891.

Singer, F.J. 1995. Effects of grazing by ungulates on upland bunchgrass communities of the northern winter range of Yellowstone National Park. Northwest Sci. 69(3):191-203.

Singer, F.J. 1996. Differences between willow communities browsed by elk and communities protected for 32 years in Yellowstone National Park. Pp. 279-290 in Effects of Grazing by Wild Ungulates in Yellowstone National Park, F.J. Singer, ed.. Tech. Rep. NPS/NRYELL/NRTR/96-01. U.S. Department of the Interior, National Park Service, Natural Resource Information Division, Denver, CO.

Singer, F.J., and R.G. Cates. 1995. Response to comment: ungulate herbivory on willows on Yellowstone's northern winter range. J. Range Manage. 48(6):563-565.

Singer, F.J., and M.K. Harter. 1996. Comparative effects of elk herbivory and 1988 fires on northern Yellowstone National Park grasslands. Ecol. Appl. 6(1):185-199.

Singer, F.J., and J.E. Norland. 1994. Niche relationships within a guild of ungulate species in Yellowstone National Park, Wyoming, following release from artificial control. Can. J. Zool. 72(8):1383-1394.

Singer, F.J., and R.A. Renkin. 1995. Effects of browsing by native ungulates on the shrubs in big sagebrush communities in Yellowstone National Park. Great Basin Nat. 55(3):201-212.

Singer, F.J., A. Harting, K.K. Symonds, and M.B. Coughenour. 1997. Density dependence, compensation, and environmental effects on elk calf mortality in Yellowstone National Park. J. Wildl. Manage. 61(1):12-25.

Singer, F.J., L.C. Mark, and R.C. Cates. 1994. Ungulate herbivory of willows on Yellowstone's northern winter range. J. Range Manage. 47(6):435-443.

Singer, F.J., W. Schreier, J. Oppenheim, and E.O. Garten. 1989. Drought, fires, and large mammals. BioScience 39(1):716-722.

Singer, F.J., K.K. Symonds, and B. Berger. 1993. Predation of Yellowstone elk calves. Park Science 13(3):18.

Singer, F.J., D.M. Swift, M.B. Coughenour, and J.D. Varley. 1998. Thunder on the Yellowstone revisited: An assessment of management of native ungulates by natural regulation, 1968-1993. Wildl. Soc. Bull. 26(3):375-390.

Singer, F.J., L.C. Zeigenfuss, and D.T. Barnett. 2000. Elk, beaver, and the persistence of willows in national parks: response to Keigley (2000). Wildl. Soc.Bull. 28(2):451-453.

Skidmore, P.B., P. Farnes, and M. Story. 1999. Hydrologic significance of Yellowstone river floods. Pp. 167-174 in The Wildland Hydrology Proceedings: Specialty Conference, June 30-July 2, 1999, Bozeman, MT, D.S. Olsen, and J.P. Potyondy, eds. Herndon, VA: American Water Resources Association.

Skinner, M.P. 1929. White-tailed deer formerly in the Yellowstone Park. J. Mammal. 10:101-115.

Smith, B.L., and R.L. Robbins. 1994. Migrations and Management of the Jackson Elk Herd. Washington, DC: U.S. Department of the Interior, National Biological Survey. 61 pp.

Smith, C.L., A.A. Simpson, and V. Bailey. 1915. Report on Investigations of the Elk Herds in the Yellowstone Region of Wyoming, Montana, and Idaho. Biological Survey and USDA Forest Service. Yellowstone National Park Library. December 14.

Smith, D.W. 1998. Yellowstone Wolf Project: Annual Report, 1997. YCR-NR-98-2. National Park Service, Yellowstone Center for Resources, Yellowstone National Park, WY.

Smith, D.W., W.G. Brewster, and E.E. Bangs. 1999a. Wolves in the Greater Yellowstone Ecosystem: restoration of a top carnivore in a complex management environment. Pp. 103-126 in Carnivores in Ecosystems: The Yellowstone Experience, T.W. Clark, A.P. Curlee, S.C. Minta, and P.M. Kareiva, eds. New Haven, CT: Yale University Press.

Smith, D.W., K.M. Murphy, and D.S. Guernsey. 1999b. Yellowstone Wolf Project:

Annual Report, 1998. YCR-NR-99-1. National Park Service, Yellowstone Center for Resources, Yellowstone National Park, Wyoming. [Online]. Available: http://www.nps.gov/yell/nature/animals/wolf/wolf99.pdf [Sept. 13, 2001].

Smith, S.G., S. Kilpatrick, A.D. Reese, B.L. Smith, T. Lemke, and D. Hunter. 1997. Wildlife habitat, feedgrounds, and brucellosis in the Greater Yellowstone Area. Pp. 65-76 in Brucellosis, Bison, Elk, and Cattle in the Greater Yellowstone Area: Defining the Problem, Exploring Solutions, E.T. Thorne, M.S. Boyce, P. Nicoletti, and T.J. Kreeger, eds. Cheyenne,WY: Wyoming Game and Fish Department.

Soether, B.E. 1997. Environmental stochasticity and population dynamics of large herbivores: A search for mechanisms. Trends Ecol. Evol. 12(4):143-149.

Soper, D. 1941. History, range, and home of the northern bison. Ecol. Monogr. 11(4):347-412.

Sprugel, D.G. 1991. Disturbance, equilibrium, and environmental variability: What is natural vegetation in a changing environment? Biol. Conserv. 58(1):1-18.

Stanley, T.R., Jr. 1995. Ecosystem management and the arrogance of humanism. Conserv. Biol. 9(2):255-262.

Stevens, M.T., M.G. Turner, G.A. Tuskan, W.H. Romme, and D.M. Waller. 1999. Genetic variation in postfire aspen seedlings in Yellowstone National Park. Mol. Ecol. 8(11): 1769-1780.

Stromberg, J.C., J. Fry, and D.T. Patten. 1997. Marsh development after large floods in an alluvial, arid-land river. Wetlands 17(2):292-300.

Stromberg, J.C., R. Tiller, and B. Richter. 1996. Effects of groundwater decline on riparian vegetation of semiarid regions: the San Pedro, Arizona. Ecol. Appl. 6(1):113-131.

Strong, D.R. 1986. Density-vague population change. Trends Ecol. Evol. 1:39-42.

Stubbs, M. 1977. Density dependence in the life-cycles of animals and its importance in K- and R- strategies. J. Animal Ecol. 46(2):677-688.

Swetnam, T.W., and J.L. Betancourt. 1998. Mesoscale disturbance and ecological response to decadal climatic variability in the American southwest. J. Climate 11(12):3128-3147.

Tahvanainen, J., E. Helle, R. Julkunen-Titto, and A. Lavola. 1985. Phenolic compounds of willow bark as deterrents against feeding by mountain hare. Oecologia 65(3):319-323.

Taper, M.L., and P.J.P. Gogan. 2002. The Nothern Yellowstone elk: density dependence and climatic conditions. J. Wildl. Manage. 66(1):106-122.

Taper, M.L., M. Meagher, and C.L. Jerde. 2000. The Phenology of Space: Spatial Aspects of Bison Density Dependence in Yellowstone National Park. Final Report from Montana State University to U.S. Geological Survey, Bozeman, MT. October 2000. 113 pp + 151 maps.

Taylor, D.L. 1973. Some ecological implications of forest fire control in Yellowstone National Park. Ecology 54(6):1394-1396.

Taylor, K.C., P.A. Mayewski, R.B. Alley, E.J. Brook, A.J. Gow, P.M. Grootes, D.A. Meese, E.S. Saltzman, J.P. Severinghaus, M.S. Twicker, J.W.C. White, S. Whitlow, and G.A. Zielinski. 1997. The Holocene-younger dryas transition recorded at Summit, Greenland. Science 278(5339):825-827.

Thorne, E.T., and J.D. Herriges, Jr. 1992. Brucellosis, wildlife and conflicts in the Greater Yellowstone Area. Trans. North Am. Wildl. Nat. Resour. Conf. 57:453-465.

Thorne, E.T., S.G. Smith, K. Aune, D. Hunter, and T.J. Roffe. 1997. Brucellosis: the disease in elk. Pp. 33-44 in Brucellosis, Bison, Elk, and Cattle in the Greater Yellowstone Area: Defining the Problem, Exploring Solutions, E.T. Thorne, M.S. Boyce, P. Nicoletti, and T.J. Kreeger, eds. Cheyenne, WY: Wyoming Game and Fish Department.

Toman, T.L., T. Lemke, L. Kuck, B.L. Smith, S.G. Smith, and K. Aune. 1997. Elk in the Greater Yellowstone Area: Status and management. Pp. 56-64 in Brucellosis, Bison, Elk, and Cattle in the Greater Yellowstone Area: Defining the Problem, Exploring Solutions, E.T. Thorne, M.S. Boyce, P. Nicoletti, and T.J. Kreeger, eds. Cheyenne,WY: Wyoming Game and Fish Department.

Tracy, B.F., and S.J. McNaughton. 1997. Elk grazing and vegetation responses following a late-season fire in Yellowstone National Park. Plant Ecol. 130(2):111-119.

Turchin, P. 1990. Rarity of density dependence or population regulation with lags? Nature 344(6267):660-663.

Turner, M.G., and W.H. Romme. 1994. Landscape dynamics in crown fire ecosystems. Landscape Ecol. 9(1):59-77.

Turner, M.G., W.H. Romme, and R.H. Gardner. 1994a. Landscape disturbance models and the long-term dynamics of natural areas. Natural Areas Journal 14(1):3-11.

Turner, M.G., Y. Wu, L.L. Wallace, W.H. Romme, and A. Brenkert. 1994b. Simulating winter interactions among ungulates vegetation, and fire in northern Yellowstone Park. Ecol. Appl. 4(3):472-496.

Turner, M.G., W.W. Hargrove, R.H. Gardner, and W.H. Romme. 1994c. Effects of fire on landscape heterogeneity in Yellowstone National Park, Wyoming. J. Veg. Sci. 5(5):731-742.

Turner, M.G., W.H. Romme, R.H. Gardner, and W.W. Hargrove. 1997. Effects of patch size and pattern on early succession on the Yellowstone National Park. Ecol. Monogr. 67(4): 411-433.

Tuskan, G.A., K.E. Francis, S.L. Russ, M.G. Turner, and W.H. Romme. 1996. RAPD markers reveal diversity within and among clonal and seedling stands of aspen in Yellowstone National Park, U.S.A. Can. J. For. Res. 26(12):2088-2098.

Tyers, D.B. 1981. The Condition of the Northern Winter Range in Yellowstone National Park: A Discussion of the Controversy. M.S. Thesis. Montana State University, Bozeman, MT. 168 pp.

Varley, N.C. 1994. Summer-fall habitat use and fall diets of mountain goats and bighorn sheep in the Absaroka Range, Montana. Pp. 131-138 in Proceedings of the 9th Biennial Symposium, Cranbrook, British Columbia, May 2-6, 1994, M. Pybus, and B. Wishart, eds. Thermopolis, WY: Northern Wild Sheep and Goat Council.

Varley, N.C.L. 1996. Ecology of Mountain Goats in the Absaroka Range, South-

Central Montana. M.S. Thesis. Montana State University, Bozeman, MT.

Wagner, F.H., R.B. Keigley, and C.L. Wambolt. 1995. Comment: Ungulate herbivory of willows on Yellowstone's northern winter range: Response to Singer et al. (1994). J. Range Manage. 48(5):475-477.

Walker, D.N. 1987. Late Pleistocene/Holocene environmental changes in Wyoming: the mammalian record. Pp. 334-392 in Late Quaternary Mammalian Biogeography and Environments of the Great Plains and Prairies, R.W. Graham, H.A. Semken, Jr., and M.A. Graham, eds. Scientific Papers Vol. 22. Springfield, IL: Illinois State Museum.

Walker, B.H., and I. Noy-Meir. 1982. Aspects of stability and resilience of savanna ecosystems. Pp. 555-590 in Ecology of Tropical Savannas, B.J. Huntley, and B.H. Walker, eds. Berlin: Springer-Verlag.

Walker, B.H., R.H. Emslie, R.N. Owensmith, and R.J. Scholes. 1987. To cull or not to cull- lessons from a Southern African drought. J. Appl. Ecol. 24(2):381-401.

Wallace, L.L., M.G. Turner, W.H. Romme, R.V. O'Neill, and Y. Wu. 1995. Scale of heterogeneity of forage production and winter foraging by elk and bison. Landscape Ecol. 10(2):75-83.

Wallmo, O.J. 1978. Mule and black-tailed deer. Pp. 31-41 in Big Game of North America: Ecology and Management, J.L. Schmidt, and D.L. Gilbert, eds. Harrisburg, PA: Stackpole Books.

Wambolt, C.L. 1996. Mule deer and elk foraging preference for 4 sagebrush taxa. J. Range Manage. 49(6):499-503.

Wambolt, C.L. 1998. Sagebrush and ungulate relationships on Yellowstones northern range. Wildl. Soc. Bull. 26(3):429-437.

Wambolt, C.L., and H.W. Sherwood. 1999. Sagebrush response to ungulate browsing in Yellowstone. J. Range Manage. 52(4):363-369.

Wambolt, C.L., M.R. Frisina, K.S. Douglass, and H.W. Sherwood. 1997. Grazing effects on nutritional quality of bluebunch wheatgrass for elk. J. Range Manage. 50(5):503-506.

Wambolt, C.L., T.L. Hoffman, and C.A. Mehus. 1999. Response of big sagebrush habitats to fire on the Northern Yellowstone Winter Range. Pp. 238-242 in Proceedings: Symposium on Shrubland Ecotones, Ephraim, UT 12-14, 1998, E.D. McArthur, W.K. Ostler, and C.L. Wambolt, eds. Proceedings RMRS-P-11. Ogden, UT: Rocky Mountain Research Station, Forest Service, U.S. Dept. of Agriculture.

Warren, E.R. 1926. A study of beaver in the Yancey region of Yellowstone National Park. Roosevelt Wildl. Annals 1:1-191.

Weaver, J.L., P.C. Paquet, and L.F. Ruggiero. 1996. Special section: Large carnivore conservation in the Rocky Mountains of the United States and Canada. Resilience and conservation of large carnivores in the Rocky Mountains. Conserv. Biol. 10(4):964-976.

Wehausen, J.D. 1996. Effects of mountain lion predation on bighorn sheep in the Sierra Nevada and Granite Mountains of California. Wildl. Soc. Bull. 24(3):471-479.

Weins, J.A., and J.T. Rotenberry. 1981. Habitat associations and community structure of birds in shrubsteppe environments. Ecol. Monogr. 51(1):21-24.

West, N.E., and M.S. Hassan. 1985. Recovery of sagebrush-grass vegetation follow-
    ing wildfire. J. Range Manage. 38(2):131-134.
Wheat, J.B. 1972. The Olsen-Chubbuck Site: A Paleo-Indian Bison Kill. Memoirs of
    the Society for American Archaeology No. 26. Washington: Society for Amer-
    ican Archaeology.
Whittaker, R.H. 1956. Vegetation of the Great Smoky Mountains. Ecol. Monogr.
    26(1):1-80.
Whittaker, R.H. 1970. Communities and Ecosystems, 2nd Ed. New York: Macmillan.
White, G.C., and R.M. Bartmann. 1997. Density dependence in deer populations. Pp.
    120-135 in The Science of Overabundance: Deer Ecology and Population Man-
    agement. W.J. McShea, H.B. Underwood, and J.H. Rappole, eds. Washington,
    DC: Smithsonian Institution Press.
White, C.A., C.E. Olmsted, and C.E. Kay. 1998. Aspen, elk and fire in the Rocky
    Mountain national parks of North America. Wildl. Soc. Bull. 26(3):449-462.
Whitlock, C. 1993. Postglacial vegetation and climate of Grand Teton and southern
    Yellowstone National Parks. Ecol. Monogr. 63(2):173-198.
Whitlock, C., P.J. Bartlein, and K.J. Van Norman. 1995. Stability of Holocene climate
    regimes in the Yellowstone region. Quat. Res. 43:433-436.
Williams, E.S., E.T. Thorne, S.L. Anderson, and J.D. Herriges. 1993. Brucellosis in free-
    ranging bison (Bison bison) from Teton County, Wyoming. J. Wildl. Dis.
    29(1):118-122.
Williams, E.S., S.L. Cain, and D.S. Davis. 1997. Brucellosis: the disease in bison. Pp.
    7-19 in Brucellosis, Bison, Elk, and Cattle in the Greater Yellowstone Area: Defin-
    ing the Problem, Exploring Solutions, E.T. Thorne, M.S. Boyce, P. Nicoletti and
    T.J. Kreeger, eds. Cheyenne, Wyoming: Wyoming Game and Fish Department.
Woodhouse, C.A., and J.T. Overpeck. 1998. 2000 years of drought variability in the
    central United States. Bull. Am. Meteorol. Soc. 79(12):2693-2714.
Wright, H.E., Jr. 1974. Landscape development, forest fires, and wilderness manage-
    ment: fire may provide the long-term stability needed to preserve certain conifer
    forest ecosystems. Science 186(4163):487-495.
Wright Jr., H.E., and M.L. Heinselman. 1973. Introduction to symposium on the
    ecological role of fire in natural coniferous forests of western and northern
    America. Quat. Res. 3:319-328.
Wright, G.H., and B.H. Thompson. 1935. Fauna of the National Parks of the United
    States. Fauna Series No. 2. Washington, DC: U.S. National Park Service. 142 pp.
Wu, Y., M.G. Turner, L.L. Wallace, and W.H. Romme. 1996. Elk survival following the
    1988 fires in Yellowstone National Park: a simulation experiment. Natural Areas
    Journal 16:198-207.
YNP (Yellowstone National Park). 1997. Yellowstone's Northern Range: Complexity
    and Change in a Wildland Ecosystem. National Park Service, Mammoth Hot
    Springs, WY.
YNP (Yellowstone National Park). 2001. Yellowstone National Park Facts. [Updated

Nov. 27, 2001]. [Online]. Available: http://www.nps.gov/yell/technical/yellfact. html [February 22, 2002].

Young, T.P. 1994. Natural die-offs of large mammals: implications for conservation. Conserv. Biol. 8(2):410-418.

Zahner, R., and N.V. DeByle. 1965. Effect of pruning the parent root on growth of aspen suckers. Ecology 46(3):373-375.

Appendix A

Understanding the Past

LATE QUATERNARY MAMMAL HISTORY OF THE GREATER YELLOWSTONE ECOSYSTEM

Mammal records from the late Pleistocene are sparse, but sites have been studied in several caves surrounding Yellowstone National Park (YNP). Usually only one species is found within one site, but these sites place the Greater Yellowstone ecosystem (GYE) in a regional perspective for the late Quaternary, especially for taxa that occur throughout the surrounding area.

Mammal faunas from these sites date from the latest Pleistocene (20,000 to 10,000 years before present [YBP]) through the Holocene (10,000 to 500 YBP). Because species responded individualistically to climate warming at the end of the Pleistocene, many of these communities do not have modern analogs. For example, most of the sites contain the remains of tundra species (pika [*Ochotona princeps*], collared lemming [*Dicrostonyx* spp.], caribou [*Rangifer tarandus*]) in association with forest mammals (porcupine [*Erethizon dorsatum*], marten [*Martes americana*], deer [*Odocoileus* spp.]) and plains-dwelling forms (bison, ground squirrels, pronghorn). A heterogeneous parkland/savanna is suggested as the best environment for supporting these types of diverse faunas (Faunmap Working Group 1996).

Extinct ungulates that appear to have inhabited the GYE include the Columbian mammoth (*Mammuthus columbi*), horse (probably several species of *Equus*), camel (*Camelops hesternus*), woodland muskox (*Bootheerium*

168

bombifrons = *Symbos cavifrons*), and mountain deer (*Navahoceros fricki*) (Faunmap Working Group 1996). In addition, several large extinct carnivores—the short-faced bear (*Arctodus simus*), American lion (*Panthera atrox*), and American cheetah (*Miracinonyx trumani*)—were common to this area (Martin and Gilbert 1978, Chomko and Gilbert 1987, Walker 1987). Other ungulates (e.g., flat-headed peccary [*Platygonus compressus*]) and carnivores (e.g., saber-tooth cat [*Smilodon floridanus*]) may also have been present, but their remains are sparse and not easily extrapolated to the GYE.

These large mammals as well as at least 24 other genera became extinct in North America at the end of the Pleistocene (11,000 YBP). The two main hypotheses about this extinction involve overexploitation by human hunters and climatically driven environmental changes (Martin and Klein 1984). Whatever the cause of the extinction, it must have had broad ramifications, including alterations in biological interactions such as predation and competition; vegetational structure and composition created by seed dispersal, browsing, and grazing; and nutrient recycling. The disappearance of species assemblages for which there is no present-day representative is coincident with the extinction event (Graham and Lundelius 1984).

As the climate began to warm about 14,000 years ago and glaciers receded northward and higher, so did some of the boreal mammalian fauna. The collared lemming that today lives on the tundra in Alaska and Canada was one of the first mammals to be extirpated from the surrounding areas and probably the GYE, having vacated the contiguous northwestern United States by at least 10,000 YBP (Faunmap Working Group 1994). Other boreal species like the pika (*O. princeps*) and the heather vole (*Phenacomys intermedius*) remained at elevations lower than their current distribution until the middle Holocene, a time of maximum warmth and dryness (Grayson 1977, 1981; Mead et al. 1982; Mead 1987). As these species dispersed to higher elevations, species like the pygmy rabbit (*Oryctolagus idahoensis*) decreased in abundance and others, which were adapted to drier habitats (e.g., *Lepus* spp.), first appeared or increased in abundance (Grayson 1987). Little is known of the Holocene history of the Great Basin ungulates (Grayson 1982, 1993).

Similar changes took place in Wyoming east of Yellowstone as summarized by Walker (1987). At 10,300 to 9,300 YBP, some boreal mammal species were still present in the basin areas, but steppe forms were starting to be found in association with boreal habitat types. However, by 5,060 to 2,760 YBP, modern mammalian distributions are believed to have been established.

Lamar Cave has yielded an extensive fossil mammal record from the

Lamar Valley for the past 3,200 years (Hadly 1996). The late Holocene fauna of the Lamar Cave is nearly identical to the modern fauna of the area. Only one species, the prairie vole (*Microtus ochrogaster*), does not occur in the modern fauna but it is found today in environments 100 km north of that site (Barnosky 1994, Hadly 1996). The sporadic presence of this species in the cave record may indicate the occurrence of more tall grass habitat in the vicinity of the cave at certain periods in the past.

Although general faunal composition has been relatively stable, relative frequencies of environmentally sensitive species fluctuated through time and appeared to correspond with climatic events in the late Holocene. High ratios of voles (*Microtus* spp.) and ground squirrels (*Spermophilus* spp.) indicate moister environments. Based on such analyses, Hadly (1996) reconstructed the environmental sequence as follows: from 2,860 to 1,370 YBP the environment was moister than it is today; from 1,500 to 560 YBP the environment became drier; cooler and moister conditions have prevailed for the last 700 years, including the Little Ice Age.

Pocket gophers (*Thomomys talpoides*) were most abundant at Lamar Cave during times with moister climates (Hadly-Barnosky 1994, Hadly 1996). Pocket gopher-body size also appears to have responded to climate change (Hadly 1997).

In other areas of the United States, similar body-size changes have been demonstrated (Purdue 1989) for ungulates such as white-tailed deer *(Odocoileus virginianus)* where body size decreased during the warmest and driest climates of the middle Holocene. Bison also became steadily smaller from the Pleistocene through the Holocene (Kurten and Anderson 1980). Although similar studies have not been conducted for the GYE, GYE ungulates probably responded similarly to middle Holocene climates.

Hadly (1996) concluded that the bones accumulated in Lamar Cave were the result of the activities of owls, small to medium-sized carnivores, and woodrats. Therefore, the faunal record from Lamar Cave is dominated by small mammals (Hadly-Barnosky 1994). Large ungulate remains (elk, deer, bison, and bighorn sheep) occur in the cave deposits but their frequencies probably do not reflect the actual abundance of those animals. Other archaeological sites in YNP (Cannon 1997) have yielded the remains of large ungulates. They document the occurrence of elk, deer, bighorn sheep, and bison in the park by about 1,200 years ago; however, they cannot be used to estimate population sizes.

Some scholars have proposed that the environments of the middle Holocene caused reductions in the size of *Bison* herds (Dillehay 1974, Frison 1978). These assumptions were based on the reduced number of *Bison* sites and their apparent absence in areas that supported *Bison* in the late and early Holocene. However, this hypothesis has not been quantitatively assessed. The variables involved must be evaluated before estimates of relative abundance over time (e.g., Cannon [1992]) can be evaluated quantitatively.

INTERPRETING BONE DEPOSITS

Factors that influence which species are likely to be preserved in particular sites are listed in Table A-1. For example, caves are frequently used as dens by predators, and consequently, the remains of predators and their prey are often preserved. At open sites, like a water hole, the remains of ungulates are far more common than the remains of predators. Also, an assemblage of bones from a pit cave, which acts as a natural trap, is quite different from that of a cave with a horizontal entrance, which permits both entrance and egress (Brain 1981).

The behavior of animals is also important in the formation of bone accumulations. Some animals may be preferentially attracted to different site types. Bats use caves as hibernacula; their remains often are very abundant in cave deposits. Predator-prey relationships may also be a critical factor in determining the composition of a bone assemblage. Owl roosts may have massive accumulations of small mammal bones but do not contain any large mammal remains. Conversely, a wolf den would primarily consist of ungulate remains.

The location of a site and its catchment area also are important. The catchment is the area from which fossils can be derived. For rodents and insectivores falling into a pit cave, the catchment may be only a few meters up to perhaps a few hundred meters across. However, the catchment of a large river system can be hundreds of square kilometers. Sites located in upland (plateaus, interfluves, etc.) and bottom land (floodplains) environments will frequently have different types of faunal assemblages. Owls may forage over a few kilometers, whereas large mammal carnivores may bring prey from tens to more than 100 kilometers. Also, distance between the site of accumulation and the site of the kill can influence what types of bones, if any, are brought to the site. This situation, which is especially important in archaeological bone assemblages, has been referred to as the "schlep effect." If an animal is killed near the site, the predator may bring the entire carcass back to the site of

TABLE A-1 Factors That Can Bias Quantitative Analyses of Bone Remains As Indices for Animal Abundance in a Biological Community

Site type[a]
Site function
Predator and prey behavior
Site location[a]
Catchment area
Schlep effect
Season of accumulation
Time averaging
Preservation of bones[a] (bone density, sediment chemistry, and depositional systems)
Methods of excavation and bone recovery[a] ability to identify skeletal elements
Amount of site excavated[a]

[a]Factors discussed by Kay (1990).

accumulation. On the other hand, if the kill is made some distance from the site of accumulation, the predator may strip the carcass of meat and not return any bones to the site of accumulation or may selectively pick bones of different nutritional value (i.e., marrow content).

Time averaging can inflate bone counts. Bones may accumulate on a surface over an extended period of time. If bones accumulate at the same rate but with greater rates of sedimentation, then bone density will be lower. The only way to correct for time averaging is to have a series of radiocarbon dates that allow sedimentation rates to be calculated.

Differential preservation of bone can also alter counts. In sediments deposited in fast currents, small and light bones can be swept away, leaving the larger and denser bones. Also, in sites in which bones are trampled or crushed by other means, foot bones are preferentially preserved, and more fragile bones like skulls and scapulas are destroyed. In an archaeological site where marrow was processed, limb bones can be reduced to small unidentifiable splinters. Kay (1990) describes these types of assemblages in his samples. Finally, bones deposited in alkaline environments like cave sediments may be preserved, but bones in acid soils (e.g., some floodplains) may dissolve over time.

The ability to identify skeletal elements to species or even generic levels varies with the taxon and the experience of the faunal analyst. Most of the skeletal elements of the ungulates should be readily identifiable, but there is a great deal of variability in the amount of bone that is identified from various sites. Also, the amount of bone from a site varies with the amount of the site excavated. Excavations to determine the potential of a site generally are limited to a relatively small volume. On other occasions, hundreds of cubic meters of a site may be removed.

Methods of collection can also bias bone samples. Semken and Graham (1996) used ordination techniques to show how species composition of archaeological sites can vary on the basis of whether the site was screened or the size of the mesh used in the screening. Sites that are not screened generally lack the bones of smaller animals and smaller bone fragments of larger species.

In most cases, bone accumulations result from multiple pathways. Also, a single site may have different stratigraphic levels that accumulated bones by different pathways. Therefore, if these levels are combined, the different pathways are mixed. To have meaningful comparisons, it is essential to compare assemblages (sites or levels within sites) with similar histories of accumulation.

A series of late Quaternary sites from the Pryor Mountains of Montana clearly illustrate the problems of various accumulation histories and how they can influence faunal samples and bone counts. The Pryor Mountains are composed of two fault-lifted blocks (East Pryor and Big Pryor) of Paleozoic sedimentary rocks. Caves are developed in the Madison Limestone on both East and Big Pryor. Two cave sites have been excavated on East Pryor that are only 300 m apart. Therefore, the caves have sampled the same environment. Also, both caves contain relatively complete sequences of Holocene deposits so they have both sampled the same time interval. The primary differences between the caves are site type and agents of bone accumulation.

One site is False Cougar Cave, which is located on East Pryor Mountain at about 8,600 ft (2,621 m). This is a small cave developed in a small outcrop of Madison Limestone. It has a horizontal entrance and contains multiple sedimentary layers that date from the late Holocene to the late Pleistocene. The cave was also used by humans, as evidenced by stone artifacts and hearths found in the cave sediments. Vegetation around the cave is a mixture of open meadow and coniferous forest.

The other cave, Shield Trap, is about 300 m west of False Cougar Cave. It is a pit cave with a vertical shaft of about 10 m. The opening to the cave is about 3 m in diameter. The cave is situated on a relatively flat ridge surrounded by open vegetation, primarily grasses. It is developed in the same limestone as False Cougar Cave.

Both caves have been excavated by the same method of 10-cm levels within natural stratigraphic units. Large specimens were piece-plotted with respect to their horizontal and vertical coordinates. In addition, the orientation and plunge of bones with long axes were measured with a Brunton compass. All sediments were collected and water screened through fine mesh (1/16-inch [0.16-cm] mesh) screens. Also, both caves contain a relatively complete sequence of Holocene sediments. Depositional environments are similar, but Shield Trap generally has more breakdown.

Therefore, the primary difference between the two sites is how each cave sampled the living fauna. Shield Trap was a pit into which animals randomly fell and were trapped. On the other hand, the horizontal entrance of False Cougar Cave did not serve as a trap. Instead, faunal remains were brought into the cave by agents, primarily predators (owls, carnivores, and humans). In some cases, bones may have accumulated as the result of an animal's dying in the cave as it was used for shelter, but these remains are an extremely minor component of the fauna.

The main difference between the bone assemblages from the two caves is in the abundance of large mammal remains. The assemblage from Shield Trap is primarily composed of bison bones, whereas False Cougar Cave contains small mammal remains (rabbit to marmot size and smaller). Shield Trap also contains small mammal remains, but they are not nearly as abundant as those from False Cougar Cave. The remains of large mammals are sparse in False Cougar Cave. Therefore, if the bone frequencies are taken at face value, Shield Trap suggests that bison were abundant throughout the Holocene. However, False Cougar Cave, which has sampled the same environment at the same time, suggests that bison, specifically, and ungulates in general were rare. Typical of most pit caves, Shield Trap also has a more complete complement of carnivores than does False Cougar Cave.

Big Lip is a rockshelter located at a lower elevation of 6,000 to 7,000 ft (1,829 to 2,134 m) on East Pryor Mountain. It contains sediments similar to those of both Shield Trap and False Cougar Cave, and it is located less than 10 miles (16 km) from these other two sites. The vegetation around the cave consists of Douglas-fir and lodgepole-pine forest. Big Lip also contains human artifacts of middle Holocene age. The primary difference between Big Lip and the other two sites is its lower elevation and vegetational surroundings. However, the fauna from Big Lip is quite different from either Shield Trap or False Cougar Cave. Big Lip rockshelter does not contain abundant microfauna, and the dominant ungulate is *Ovis canadensis*. Based on the artifact assemblage and bone breakage patterns, it appears that Big Lip served as a hunting and butchering site for bighorn sheep. Again, Big Lip would provide a very different picture of ungulate abundance if faunal remains were interpreted at face value without considering accumulation pathways.

These problems apply to the sites analyzed and interpreted by Kay (1990). For the Myers-Hindman site, Kay (1990) lumped faunal remains from seven cultural levels and eight settlement units dating from 9,000 to 800 YBP and calculated total Minimum Number of Individuals (MNI) for the site. He then compared taxa. This technique averages and smears any fluctuations in abundance due to environmental fluctuations throughout the entire Holocene. Also, the percentage comparisons between taxa are not independent. Kay (1990) concluded that the proportions of ungulates did not correspond to today's relative abundance of ungulate species in the vicinity of the site.

For Mummy Cave, Kay (1990) again lumped MNI and Number of Individual Specimens (NISP) values for 38 distinct layers that date between 9,000 and 300 YBP. Also, he noted that the faunal remains from the site have never been completely identified. As indicated by Kay (1990), the materials reported by Harris (1978) represent only a sample of the entire site. Again, by comparing the lumped MNI values for the entire site for the various ungulates, Kay (1990) concluded that these proportions were quite different from the proportions of the ungulate species in modern populations.

For the Dead Indian Creek site, Kay did not explicitly explain how he treated the sample, but again it appears that he calculated MNIs for the various ungulate species for the entire site and then compared proportions. These proportions of ungulates did not match those of modern populations.

The Bugas-Holding site is a single-component and probably single-occupation site, which is an ideal sample. However, again the proportions of ungulates in the faunal sample did not match those of the modern populations. However, Kay (1990) did not consider factors other than actual abundances that could have caused this difference. The sample is dominated by bison. One possibility for this bias is in the methods of procurement of bison and elk. Frison (1974) notes that there are many close parallels between handling bison and domestic cattle. He further suggests that there may be a critical size for a manageable herd that would then result in the slaughter of more animals (Frison 1974). In comparison, there is no known evidence for artificial elk traps or communal procurement as with bison (Frison 1978). However, if the mature female leader is killed the remainder of the elk herd often mill in circles, and the entire herd, or a good portion of it, can be killed. On the other hand, the herd can disperse for several miles and be hard to find (Frison 1978). These differences, as well as others, could easily account for the faunal differences at Bugas-Holding and hence the abundance of remains at the site is not necessarily related to the abundance of the animals.

The Joe Miller site (48AB18) is the only archaeological excavation in Wyoming where elk are the most common ungulate (Kay 1990). Creasman et al. (1982) concluded that the upper component of the Joe Miller site represents an area for processing elk.

Kay has shown that elk bones are generally not as abundant as the bones of other ungulates such as bison, deer, sheep, and pronghorn. The paucity of elk bones could reflect low population levels as postulated by Kay (1990) and others (Keigley and Wagner 1998), but it is just as likely, and perhaps more probable, that differences in the abundances of bones are an artifact of processes by which bones were accumulated. Also, lumping of quantitative data from different stratigraphic levels, as done by Kay (1990), has created samples time-averaged over as much as 10,000 years, virtually the span of the entire Holocene. Also, grouping of data, as done by Kay (1990) by combining different stratigraphic units within sites, probably mixes various accumulation pathways for these stratigraphic levels, which again would significantly bias the frequency distributions.

Appendix B

Biographical Information on The Committee on Ungulate Management in Yellowstone National Park

David R. Klein (Chair) is Emeritus Professor, Institute of Arctic Biology and Department of Biology and Wildlife at the University of Alaska Fairbanks. He earned his B.S. in zoology/wildlife at the University of Connecticut, M.S. in wildlife management from the University of Alaska, and Ph.D. in zoology/ecology at the University of British Columbia. Dr. Klein's research interests include ungulate ecology, with emphasis on forage relationships, and land-use policy and resource management in the north. He was a member of the NRC Committee on Management of Wolf and Bear Populations in Alaska.

Dale R. McCullough (Vice Chair) is Professor of Wildlife Biology in the Ecosystem Sciences Division of the Department of Environmental Science, Policy, and Management and Resource Conservation in the Museum of Vertebrate Zoology at the University of California, Berkeley, where he holds the A. Starker Leopold endowed chair. He received his B.S. in wildlife management from South Dakota State University, M.S. in wildlife management from Oregon State University, and Ph.D. in zoology from the University of California, Berkeley. His research interests concern the behavior, ecology, conservation, and management of large mammals. Dr. McCullough has served on four NAS/NRC committees reviewing wildlife issues, most recently as Principal Investigator on the study Brucellosis in the Greater Yellowstone Area.

Barbara H. Allen-Diaz is Professor of Environmental Science, Policy, and Management at the University of California, Berkeley. She received her A.B. in anthropology, M.S. in range management, and Ph.D. in wildland resource science from the University of California, Berkeley. Her research interests include plant community succession and classification, meadow, hardwood rangeland ecology, forest grazing, hydrology, and range management.

Norman F. Cheville is Dean of the College of Veterinary Medicine at Iowa State University. He received a D.V.M. from Iowa State University and M.S. and Ph.D. from the University of Wisconsin. In 1968, he served a sabbatical year at the National Institute for Medical Research, London, studying under Anthony Allison. The honorary degree *Doctor Honoris Causa* was conferred by the University of Liége in 1986 for outstanding work in veterinary pathology. Dr. Cheville served as Principal Investigator on the NRC study Brucellosis in the Greater Yellowstone Area.

Russell W. Graham is Chief Curator at the Denver Museum of Nature and Science. He received his Ph.D. from the University of Texas at Austin. His research interests include the evolution and biogeography of Quaternary mammal communities. He currently serves on the NRC's U.S. National Committee for the International Union for Quaternary Research and the National Committee for DIVERSITAS.

John E. Gross is Senior Research Scientist with the Division of Sustainable Ecosystems, Commonwealth Scientific and Industrial Research Organisation, Australia, and Research Associate with the Natural Resource Ecology Laboratory, Colorado State University. He earned his B.A. from the University of Colorado in biology, M.S. in zoology from Colorado State University, and Ph.D. in ecology from the University of California, Davis. Dr. Gross's research interests include conservation biology, ecological modeling, and the population and nutritional ecology of herbivores.

James A. MacMahon is Trustee Professor of Biology at Utah State University. He earned his B.S. in zoology at Michigan State University and Ph.D. in biology at the University of Notre Dame. His research interests include theory of community organization, community ecology of deserts, biology of desert perennials, energy exchange in plant and animal populations, biology of reptiles and amphibians, and biology of arachnids. Dr. MacMahon is currently a member of the Board on Environmental Studies and Toxicology and the Committee on Future Roles, Challenges, and Opportunities for the U.S. Geological Survey.

Nancy E. Mathews is Assistant Professor, Department of Wildlife Ecology at the University of Wisconsin, Madison. Her research interests include behavioral ecology, conservation biology, and large-scale assessments of biodiversity. Dr. Mathews received her B.S. in biology from Pennsylvania State University, M.S. in forest biology (wildlife management) and Ph.D. in forest biology (ecology) from the State University of New York, College of Environmental Science and Forestry (SUNY-CESF).

Duncan T. Patten is Research Professor in the Big Sky Institute at Montana State University, Bozeman, Montana. He is also Professor Emeritus of Plant Biology and past director of the Center for Environmental Studies at Arizona State University. Dr. Patten received an A.B. from Amherst College, M.S. from the University of Massachusetts at Amherst, and Ph.D. from Duke University. His research interests include arid and mountain ecosystems, especially the understanding of ecological processes and restoration of western riparian and wetland ecosystems. He has been a member of the NRC's Commission on Geoscience, Environment and Resources, the Board on Environmental Studies, and numerous NAS/NRC committees.

Katherine Ralls is a research zoologist at the Smithsonian Institution, Washington, DC. She has a B.A. from Stanford University, an M.A. from Radcliffe College, and a Ph.D. in biology from Harvard University. Her areas of expertise are the biology of mammals, mammalian behavior, conservation biology, the genetic problems of small captive and wild populations, field studies of threatened and endangered species, and the development and testing of decision-making tools to improve management of threatened and endangered species. Dr. Ralls served previously on the NRC's Committee on Scientific Issues in the Endangered Species Act.

Monica G. Turner is Professor of Terrestrial Ecology in the Department of

Zoology at the University of Wisconsin, Madison. She received her Ph.D. in ecology from the University of Georgia. Her research interests include landscape ecology, ecological modeling, and natural disturbance dynamics. Currently, she serves on the NRC's Ecosystems Panel. Previously, she served on the NRC's Committee on Scientific and Technical Criteria for Federal Acquisition of Lands for Conservation.

Elizabeth S. Williams is Professor, Department of Veterinary Sciences, University of Wyoming, and a veterinary pathologist at the Wyoming State Veterinary Laboratory. She received a BS in zoology from the University of Maryland at College Park, D.V.M. in veterinary medicine from Purdue University, and Ph.D. in veterinary pathology from Colorado State University.